CHEMOMETRICS IN ANALYTICAL SPECTROSCOPY

RSC Analytical Spectroscopy Monographs

Series Editor: Neil Barnett, *Deakin University, Victoria, Australia.*

Advisory Panel: F. Adams, *Universitaire Instelling Antwerp, Wirijk, Belgium*; R. Browner, *Georgia Institute of Technology, Georgia, USA*; J. Callis, *Washington University, Washington, USA*; J. Chalmers, *ICI Materials, Middlesbrough, UK*; J. Monaghan, *ICI Chemicals and Polymers Ltd., Runcorn, UK*; A. Sanz Medel, *Universidad de Oviedo, Spain*; R. Snook, *UMIST, Manchester, UK.*

The series aims to provide a tutorial approach to the use of spectrometric and spectroscopic measurement techniques in analytical science, providing guidance and advice to individuals on a day-to-day basis during the course of their work with the emphasis on important practical aspects of the subject.

Flame Spectrometry in Environmental Chemical Analysis: A Practical Guide by Malcolm S. Cresser, *Department of Plant and Soil Science, University of Aberdeen, Aberdeen, UK*

Chemometrics in Analytical Spectroscopy by Mike J. Adams, *School of Applied Sciences, University of Wolverhampton, Wolverhampton, UK*

How to obtain future titles on publication

A standing order plan is available for this series. A standing order will bring delivery of each new volume immediately upon publication. For further information, please write to:

Turpin Distribution Services Ltd.
Blackhorse Road
Letchworth
Herts. SG6 1HN

Telephone: Letchworth (01462) 672555

RSC
ANALYTICAL
SPECTROSCOPY
MONOGRAPHS

Chemometrics in Analytical Spectroscopy

Mike J. Adams

School of Applied Sciences, University of Wolverhampton, Wolverhampton, UK

WITHDRAWN

THE ROYAL
SOCIETY OF
CHEMISTRY

A catalogue record for this book is available from the British Library.

ISBN 0-85404-555-4

Published by The Royal Society of Chemistry,
Thomas Graham House, Science Park, Cambridge CB4 4WF

Typeset by Computape (Pickering) Ltd, Pickering, North Yorkshire
Printed by Bookcraft (Bath) Ltd.

Preface

The term chemometrics was proposed more than 20 years ago to describe the techniques and operations associated with the mathematical manipulation and interpretation of chemical data. It is within the past 10 years, however, that chemometrics has come to the fore, and become generally recognized as a subject to be studied and researched by all chemists employing numerical data. This is particularly true in analytical science. In a modern instrumentation laboratory, the analytical chemist may be faced with a seemingly overwhelming amount of numerical and graphical data. The identification, classification and interpretation of these data can be a limiting factor in the efficient and effective operation of the laboratory. Increasingly, sophisticated analytical instrumentation is also being employed out of the laboratory, for direct on-line or in-line process monitoring. This trend places severe demands on data manipulation, and can benefit from computerized decision making.

Chemometrics is complementary to laboratory automation. Just as automation is largely concerned with the tools with which to handle the mechanics and chemistry of laboratory manipulations and processes, so chemometrics seeks to apply mathematical and statistical operations to aid data handling.

This book aims to provide students and practising spectroscopists with an introduction and guide to the application of selected chemometric techniques used in processing and interpreting analytical data. Chapter 1 covers the basic elements of univariate and multivariate data analysis, with particular emphasis on the normal distribution. The acquisition of digital data and signal enhancement by filtering and smoothing are discussed in Chapter 2. These processes are fundamental to data analysis but are often neglected in chemometrics research texts. Having acquired data, it is often necessary to process them prior to analysis. Feature selection and extraction are reviewed in Chapter 3; the main emphasis is on deriving information from data by forming linear combinations of measured variables, particularly principal components. Pattern recognition comprises a wide variety of chemometric and multivariate statistical techniques and the most common algorithms are described in Chapters 4 and 5. In Chapter 4, exploratory data analysis by clustering is discussed, whilst Chapter 5 is concerned with classification and discriminant analysis. Multivariate calibration techniques have become increasingly popular and Chapter 6 provides a summary and examples of the more common algorithms in use. Finally, an Appendix is included which aims to serve as an introduction or refresher in matrix algebra.

A conscious decision has been made not to provide computer programs of the algorithms discussed. In recent years, the range and quality of software available commercially for desktop, personal computers has improved dramatically. Statistical software packages with excellent graphic display facilities are available from many sources. In addition, modern mathematical software tools allow the user to develop and experiment with algorithms without the problems associated with developing machine specific input/output routines or high resolution graphic interfaces.

The text is not intended to be an exhaustive review of chemometrics in spectroscopic analysis. It aims to provide the reader with sufficient detail of fundamental techniques to encourage further study and exploration, and aid in dispelling the 'black-box' attitude to much of the software currently employed in instrumental analytical analysis.

Contents

CHAPTER 1

Descriptive Statistics

1 Introduction

The mathematical manipulation of experimental data is a basic operation associated with all modern instrumental analytical techniques. Computerization is ubiquitous and the range of computer software available to spectroscopists can appear overwhelming. Whether the final result is the determination of the composition of a sample or the qualitative identification of some species present, it is necessary for analysts to appreciate how their data are obtained and how they can be subsequently modified and transformed to generate the required information. A good starting point in this understanding is the study of the elements of statistics pertaining to measurement and errors.[1–3] Whilst there is no shortage of excellent books on statistics and their applications in spectroscopic analysis, no apology is necessary here for the basics to be reviewed.

Even in those cases where an analysis is qualitative, quantitative measures are employed in the processes associated with signal acquisition, data extraction, and data processing. The comparison of, say, a sample's infrared spectrum with a set of standard spectra contained in a pre-recorded database involves some quantitative measure of similarity in order to find and identify the best match. Differences in spectrometer performance, sample preparation methods, and the variability in sample composition due to impurities will all serve to make an exact match extremely unlikely. In quantitative analysis the variability in results may be even more evident. Within-laboratory tests amongst staff and inter-laboratory round-robin exercises often demonstrate the far from perfect nature of practical quantitative analysis. These experiments serve to confirm the need for analysts to appreciate the source of observed differences and to understand how such errors can be treated to obtain meaningful conclusions from the analysis.

Quantitative analytical measurements are always subject to some degree of

[1] C. Chatfield, 'Statistics for Technology', Chapman and Hall, London, UK, 1976.
[2] P.R. Bevington, 'Data Reduction and Error Analysis for the Physical Sciences', McGraw-Hill, New York, USA, 1969.
[3] J.C. Miller and J.N. Miller, 'Statistics for Analytical Chemistry', Ellis Horwood, Chichester, UK, 1993.

error. No matter how much care is taken, or how stringent the precautions followed to minimize the effects of gross errors from sample contamination or systematic errors from poor instrument calibration, random errors will always exist. In practice this means that although a quantitative measure of any variable, be it mass, concentration, absorbance value, *etc.*, may be assumed to approximate the unknown true value, it is unlikely to be exactly equal to it. Repeated measurement of the same variable on similar samples will not only provide discrepancies between the observed results and the true value, but there will be differences between the measurements themselves. This variability can be ascribed to the presence of random errors associated with the measurement process, *e.g.* instrument generated noise, as well as the natural, random variation in any sample's characteristics and composition. As more samples are analysed or more measurements are repeated then a pattern to the inherent scatter of the data will emerge. Some values will be observed to be too high and some too low compared with the correct result, if this is known. In the absence of any bias or systematic error the results will be distributed evenly about the true value. If the analytical process and repeating measurement exercise could be undertaken indefinitely, then the true underlying distribution of the data about the correct or expected value would be obtained. In practice, of course, this complete exercise is not possible. It is necessary to hypothesize about the scatter of observed results and assume the presence of some underlying predictable and well characterized parent distribution. The most common assumption is that the data are distributed *normally*.

2   Normal Distribution

The majority of statistical tests, and those most widely employed in analytical science, assume that observed data follow a normal distribution. The normal, sometimes referred to as *Gaussian*, distribution function is the most important distribution for continuous data because of its wide range of practical application. Most measurements of physical characteristics, with their associated random errors and natural variations, can be approximated by the normal distribution. The well known shape of this function is illustrated in Figure 1. As shown, it is referred to as the normal probability curve.[2] The mathematical model describing the normal distribution function with a single measured variable, *x*, is given by Equation (1).

$$f(x) = \frac{1}{\sigma\sqrt{(2\pi)}} \exp\left[\frac{-(x-\mu)^2}{2\sigma^2}\right] \tag{1}$$

The height of the curve at some value of *x* is denoted by $f(x)$ while μ and σ are characteristic parameters of the function. The curve is symmetric about μ, the *mean* or average value, and the spread about this value is given by the *variance*, σ^2, or *standard deviation*, σ. It is common for the curve to be standardized so that the area enclosed is equal to unity, in which case $f(x)$ provides the probability of observing a value within a specified range of *x* values. With

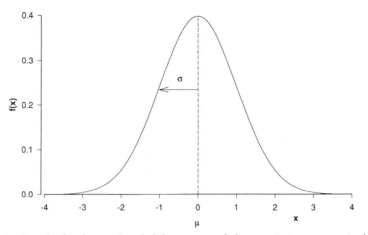

Figure 1 *Standardized normal probability curve and characteristic parameters, the mean and standard deviation*

reference to Figure 1, one-half of observed results can be expected to lie above the mean and one-half below μ. Whatever the values of μ and σ, about one result in three will be expected to be more than one standard deviation from the mean, about one in twenty will be more than two standard deviations from the mean, and less than one in 300 will be more than 3σ from μ.

Equation (1) describes the idealized distribution function, obtained from an infinite number of sample measurements, the so-called *parent population distribution*. In practice we are limited to some finite number, n, of samples taken from the population being examined and the *statistics*, or estimates, of mean, variance, and standard deviation are denoted then by $\bar{x}$, s^2, and s respectively. The mathematical definitions for these parameters are given by equations (2)–(4),

$$\bar{x} = \sum_{i=1}^{n} x_i/n \tag{2}$$

$$s^2 = \sum_{i=1}^{n} (x_i - \bar{x})^2/(n-1) \tag{3}$$

$$s = \sqrt{(s^2)} \tag{4}$$

where the subscript i ($i = 1 \ldots n$) denotes the individual elements of the set of data.

A simple example serves to illustrate the use of these statistics in reducing data to key statistical values. Table 1 gives one day's typical laboratory results for 40 mineral water samples analysed for sodium content by flame photometry. In analytical science it is common practice for such a list of replicated analyses to be reduced to these descriptive statistics. Despite their widespread use and analysts' familiarity with these elementary statistics care must be taken with

Table 1   *The sodium content* (mg kg^{-1}) *of bottled mineral water as determined by flame photometry*

			Sodium (mg kg^{-1})		
	10.8	10.4	11.7	10.6	12.2
	11.1	12.2	11.3	11.5	10.2
	10.6	11.6	10.2	11.2	10.6
	10.9	10.2	10.3	10.2	10.3
	11.5	10.6	10.5	10.2	10.1
	11.2	12.4	12.4	10.4	12.5
	10.5	11.6	10.3	10.5	11.6
	11.8	12.3	10.1	12.2	10.8
Group Means:	11.1	11.4	10.8	10.8	11.0
Group s^2:	0.197	0.801	0.720	0.514	0.883
Group s:	0.444	0.895	0.848	0.717	0.939

Total Mean = 11.04 mg kg^{-1}
$s^2 =$ 0.60 mg^2 kg^{-2}
$s =$ 0.78 mg kg^{-1}
%*RSD* = 7.03%

Table 2   *Concentration of chromium and nickel, determined by AAS, in samples taken from four sources of waste waters*

Source	*A*		*B*		*C*		*D*		
	Cr	Ni	Cr	Ni	Cr	Ni	Cr	Ni	mg kg^{-1}
	10	8.04	10	9.14	10	7.46	8	6.58	
	8	6.95	8	8.14	8	6.77	8	5.76	
	13	7.58	13	8.74	13	12.74	8	7.71	
	9	8.81	9	8.77	9	7.11	8	8.84	
	11	8.33	11	9.26	11	7.81	8	8.47	
	14	9.96	14	8.10	14	8.84	8	7.04	
	6	7.24	6	6.13	6	6.08	8	5.25	
	4	4.26	4	3.10	4	5.39	19	12.5	
	12	10.84	12	9.13	12	8.15	8	5.56	
	7	4.82	7	7.26	7	6.42	8	7.91	
	5	5.68	5	4.74	5	5.74	8	6.90	
Mean	9	7.50	9	7.50	9	7.50	9	7.50	
s	3.16	1.94	3.16	1.94	3.16	1.94	3.16	1.94	
r		0.82		0.82		0.82		0.82	

their application and interpretation; in particular, what underlying assumptions have been made. In Table 2 is a somewhat extreme but illustrative set of data. Chromium and nickel concentrations have been determined in waste water supplies from four different sources (A, B, C and D). In all cases the mean concentration and standard deviation for each element is similar, but careful examination of the original data shows major differences in the results and element distribution. These data will be examined in detail later. The practical significance of reducing the original data to summary statistics is questionable and may serve only to hide rather than extract information. As a general rule, it is always a good idea to examine data carefully before and after any transformation or manipulation to check for absurdities and loss of information.

Whilst both variance and standard deviation attempt to describe the width of the distribution profile of the data about a mean value, the standard deviation is often favoured over variance in laboratory reports as s is expressed in the same units as the original measurements. Even so, the significance of a standard deviation value is not always immediately apparent from a single set of data. Obviously a large standard deviation indicates that the data are scattered widely about the mean value and, conversely, a small standard deviation is characteristic of a more tightly grouped set of data. The terms 'large' and 'small' as applied to standard deviation values are somewhat subjective, however, and from a single value for s it is not immediately apparent just how extensive the scatter of values is about the mean. Thus, although standard deviation values are useful for comparing sets of data, a further derived function, usually referred to as the *relative standard deviation*, RSD, or *co-efficient of variation*, CV, is often used to express the distribution and spread of data.

$$\%\text{CV}, \%\text{RSD} = 100s/\bar{x} \qquad (5)$$

If sets or groups of data of equal size are taken from the parent population then the mean of each group will vary from group to group and these mean values form the sampling distribution of $\bar{x}$. As an example, if the analytical results provided in Table 1 are divided into five groups, each of eight results, then the group mean values are 11.05, 11.41, 10.85, 10.85, and 11.04 mg kg^{-1}. The mean of these values is still 11.04, but the standard deviation of the group means is 0.23 mg kg^{-1} compared with 0.78 mg kg^{-1} for the original 40 observations. The group means are less widely scattered about the mean than the original data (Figure 2). The standard deviation of group mean values is referred to as the *standard error of the sample mean*, σ_m, and is calculated from

$$\sigma_m = \sigma_p/\sqrt{n} \qquad (6)$$

where σ_p is the standard deviation of the parent population and n is the number of observations in each group. It is evident from Equation (6) that the more observations taken, then the smaller the standard error of the mean and the more accurate the value of the mean. This distribution of sampled mean values provides the basis for an important concept in statistics. If random samples of group size n are taken from a normal distribution then the distribution of the sample means will also be normal. Furthermore, and this is not intuitively obvious, even if the parent distribution is not normal, providing large sample sizes ($n > 30$) are taken then the sampling distribution of the group means will

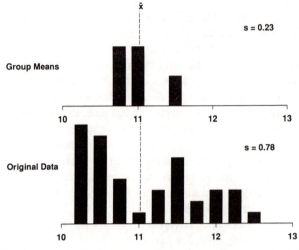

Figure 2 *Group means for the data from Table 1 have a lower standard deviation than the original data*

still approximate the normal curve. Statistical tests based on an assumed normal distribution can therefore be applied to essentially non-normal data. This result is known as the *central limit theorem* and serves to emphasize the importance and applicability of the normal distribution function in statistical data analysis since non-normal data can be normalized and can be subject to statistical analysis.[1,3]

Significance Tests

Having introduced the normal distribution and discussed its basic properties, we can move on to the common statistical tests for comparing sets of data. These methods and the calculations performed are referred to as *significance tests*. An important feature and use of the normal distribution function is that it enables areas under the curve, within any specified range, to be accurately calculated. The function in Equation (1) is integrated numerically and the results presented in statistical tables as areas under the normal curve. From these tables, approximately 68% of observations can be expected to lie in the region bounded by one standard deviation from the mean, 95% within $\mu \pm 2\sigma$, and more than 99% within $\mu \pm 3\sigma$.

We can return to the data presented in Table 1 for the analysis of the mineral water. If the parent population parameters, σ and μ_0, are known to be 0.82 mg kg^{-1} and 10.8 mg kg^{-1} respectively, then can we answer the question of whether the analytical results given in Table 1 are likely to have come from a water sample with a mean sodium level similar to that providing the parent data. In statistic's terminology, we wish to test the *null hypothesis* that the means of the sample and the suggested parent population are similar. This is generally written as

$$H_0: \bar{x} = \mu_0 \tag{7}$$

i.e. there is no difference between x and μ_0 other than that due to random variation. The lower the probability that the difference occurs by chance, the less likely it is that the null hypothesis is true. In order for us to make the decision whether to accept or reject the null hypothesis, we must declare a value for the chance of making the wrong decision. If we assume there is less than a 1 in 20 chance of the difference being due to random factors, the difference is *significant* at the 5% level (usually written as $\alpha = 5\%$). We are willing to accept a 5% risk of rejecting the conclusion that the observations are from the same source as the parent data if they are in fact similar.

The test statistic for such an analysis is denoted by z and is given by

$$z = \frac{\bar{x} - \mu_0}{\sigma/\sqrt{n}} \tag{8}$$

$\bar{x}$ is 11.04 mg kg^{-1}, as determined above, and substituting into Equation (8) values for μ_0 and σ then

$$z = \frac{11.04 - 10.80}{0.82/\sqrt{40}} = 1.85 \tag{9}$$

The extreme regions of the normal curve containing 5% of the area are illustrated in Figure 3 and the values can be obtained from statistical tables. The selected portion of the curve, dictated by our limit of significance, is referred to as the critical region. If the value of the test statistic falls within this area then the hypothesis is rejected and there is no evidence to suggest that the samples come from the parent source. From statistic tables, 2.5% of the area is below -1.96σ and 97.5% is above 1.96σ. The calculated value for z of 1.85 does not exceed the tabulated z-value of 1.96 and the conclusion is that the mean sodium concentrations of the analysed samples and the known parent sample are not significantly different.

In the above example it was assumed that the mean value and standard deviation of the sodium concentration in the parent sample were known. In practice this is rarely possible as all the mineral water from the source would not have been analysed and the best that can be achieved is to obtain recorded estimates of μ and σ from repetitive sampling. Both the recorded mean value and the standard deviation will undoubtedly vary and there will be a degree of uncertainty in the precise shape of the parent normal distribution curve. This uncertainty, arising from the use of sampled data, can be compensated for by using a probability distribution with a wider spread than the normal curve. The most common such distribution used in practice is *Student's t-distribution*. The *t*-distribution curve is of a similar form to the normal function. As the number of samples selected and analysed increases the two functions become increasingly more similar.[1] Using the *t*-distribution the well known *t*-test can be performed to establish the likelihood that a given sample is a member of a

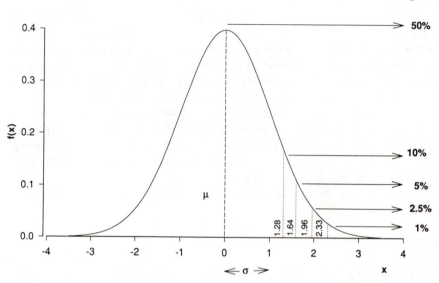

Figure 3 *Areas under the normal curve and z values for some critical regions*

population with specified characteristics. Replacing the *z*-statistic by the *t*-statistic implies that we must specify not only the level of significance, α, of the test, but also the so-called number of *degrees of freedom, i.e.* the number of independent measures contributing to the set of data. From the data supplied in Table 1, is it likely that these samples of mineral water came from a source with a mean sodium concentration of more than 10.5 mg kg^{-1}?

Assuming the samples were randomly collected, then the *t*-statistic is computed from

$$t = \frac{\bar{x} - \mu_0}{s/\sqrt{n}} \tag{10}$$

where $\bar{x}$ and *s* are our calculated estimates of the sample mean and standard deviation, respectively. From standard tables, for 39 degrees of freedom, $n - 1$, and with a 5% level of significance the value of *t* is given as 1.68. From Equation (10), $t = 4.38$ which exceeds the tabulated value of *t* and thus lies in the critical region of the *t*-curve. Our conclusion is that the samples are unlikely to arise from a source with a mean sodium level of 10.5 mg kg^{-1} or less, leaving the alternative hypothesis that the sodium concentration of the parent source is greater than this.

The *t*-test can also be employed in comparing statistics from two different samples or analytical methods rather than comparing, as above, one sample against a parent population. The calculation is only a little more elaborate, involving the standard deviation of two data sets to be used. Suppose the results from the analysis of a second day's batch of 40 samples of water give a mean value of 10.9 mg kg^{-1} and standard deviation of 0.83 mg kg^{-1}. Are the mean sodium levels from this set and the data in Table 1 similar, and could the samples come from the same parent population?

For this example the *t*-test takes the form

$$t = \frac{\bar{x}_1 - \bar{x}_2}{s_p \sqrt{(1/n_1 + 1/n_2)}} \tag{11}$$

The quantity s_p is the *pooled estimate* of the parent population standard deviation and, for equal numbers of samples in the two sets ($n_1 = n_2$), is given by

$$s_p^2 = (s_1^2 + s_2^2)/2 \tag{12}$$

where s_1 and s_2 are the standard deviations for the two sets of data.

Substituting the experimental values in Equations (11) and (12) provides a *t*-value of 0.78. Accepting once again a 5% level of significance, the tabulated value of *t* for 38 degrees of freedom and $\alpha = 0.025$ is 2.02. (Since the mean of one set of data could be significantly higher or lower than the other, an α value of 2.5% is chosen to give a combined 5% critical region, a so-called *two-tailed application*.) As the calculated *t*-value is less than the tabulated value then there is no evidence to suggest that the samples came from populations having different means. Hence, we accept that the samples are similar.

The *t*-test is widely used in analytical laboratories for comparing samples and methods of analysis. Its application, however, relies on three basic assumptions. Firstly, it is assumed that the samples analysed are selected at random. This condition is met in most cases by careful design of the sampling procedure. The second assumption is that the parent populations from which the samples are taken are normally distributed. Fortunately, departure from normality rarely causes serious problems providing sufficient samples are analysed. Finally, the third assumption is that the population variances are equal. If this last criterion is not valid then errors may arise in applying the *t*-test and this assumption should be checked before other tests are applied. The equality of variances can be examined by application of the *F-test*.

The *F*-test is based on the *F*-probability distribution curve and is used to test the equality of variances obtained by statistical sampling. The distribution describes the probabilities of obtaining specified ratios of sample variance from the same parent population. Starting with a normal distribution with variance σ^2, if two random samples of sizes n_1 and n_2 are taken from this population and the sample variances, s_1^2 and s_2^2, calculated then the quotient s_1^2/s_2^2 will be close to unity if the sample sizes are large. By taking repeated pairs of samples and plotting the ratio, $F = s_1^2/s_2^2$, the *F*-distribution curve is obtained.[3]

In comparing sample variances, the ratio s_1^2/s_2^2 for the two sets of data is computed and the probability assessed, from *F*-tables, of obtaining by chance that specific value of *F* from two samples arising from a single normal population. If it is unlikely that this ratio could be obtained by chance, then this is taken as indicating that the samples arise from different parent populations with different variances.

A simple application of the *F*-test can be illustrated by examining the mineral water data in the previous examples for equality of variance.

The *F*-ratio is given by,

$$F = s_1^2/s_2^2 \qquad (13)$$

which for the experimental data gives $F = 1.064$.

Each variance has 39 degrees of freedom ($n - 1$) associated with it, and from tables, the *F*-value at the 5% confidence level is approximately 1.80. The *F*-value for the experimental data is less than this and, therefore, does not lie in the critical region. Hence, the hypothesis that the two samples came from populations with similar variances is accepted.

In preceeding examples we have been comparing distributions of variates measured in the same units, *e.g.* mg kg^{-1}. Comparing variates of differing units can be achieved by transforming our data by the process called *standardization* which results in new values that have a mean of zero and unit standard deviation. Standardization is achieved by subtracting a variable's mean value from each individual value and dividing by the standard deviation of the variable's distribution. Denoting the new standardized value as z, gives

$$z = x_i - \bar{x}/s \qquad (14)$$

Standardization is a common transformation procedure in statistics and chemometrics. It should be used with care as it can distort data by masking major differences in relative magnitudes between variables.

Analysis of Variance

The tests and examples discussed above have concentrated on the statistics associated with a single variable and comparing two samples. When more samples are involved a new set of techniques is used, the principal methods being concerned with the analysis of variance. Analysis of variance plays a major role in statistical data analysis and many texts are devoted to the subject.[3-8] Here, we will only discuss the topic briefly and illustrate its use in a simple example.

Consider an agricultural trial site sampled to provide six soil samples which are subsequently analysed colorimetrically for phosphate concentration. The task is to decide whether the phosphate content is the same in each sample.

[4] H.L. Youmans, 'Statistics for Chemists', J. Wiley, New York, USA, 1973.
[5] G.E.P. Box, W.G. Hunter, and J.S. Hunter, 'Statistics for Experimenters', J. Wiley, New York, USA, 1978.
[6] D.L. Massart, A. Dijkstra, and L. Kaufman, 'Evaluation and Optimisation of Laboratory Methods and Analytical Procedures', Elsevier, London, UK, 1978.
[7] L. Davies, 'Efficiency in Research, Development and Production: The Statistical Design and Analysis of Chemical Experiments', The Royal Society of Chemistry, Cambridge, UK, 1993.
[8] M.J. Adams, in 'Practical Guide to Chemometrics', ed. S.J. Haswell, Marcel Dekker, New York, USA, 1992, p. 181.

Table 3 *Concentration of phosphate* (mg kg^{-1}), *determined colorimetrically, in five sub-samples of soils from six field sites*

	Phosphate (mg kg^{-1})					
Sample	1	2	3	4	5	6
Sub-sample						
i	51	49	56	56	48	56
ii	54	56	58	48	51	52
iii	53	51	52	52	57	52
iv	48	49	51	58	55	58
v	47	48	58	51	53	56

Table 4 *Commonly used table layout for the analysis of variance (ANOVA) and calculation of the F-value statistic*

Source of variation	Sum of squares	Degrees of freedom	Mean squares	F-Test
Among samples	SS_A	$m-1$	s_A^2	s_A^2/s_W^2
Within samples	SS_W	$N-m$	s_W^2	
Total variation	SS_T	$N-1$	s_T^2	

A common problem with this type of data analysis is the need to separate the *within-sample variance, i.e.* the variation due to sample inhomogeneity and analytical errors, from the variance which exists due to differences between the phosphate content in the samples. The experimental procedure is likely to proceed by dividing each sample into sub-samples and determining the phosphate concentration of each sub-sample. This process of analytical replication serves to provide a means of assessing the within-sample variations due to experimental error. If this is observed to be large compared with the variance between the samples it will obviously be difficult to detect the differences between the six samples. To reduce the chance of introducing a systematic error or bias in the analysis, the sub-samples are randomized. In practice, this means that the sub-samples from all six samples are analysed in a random order and the experimental errors are *confounded* over all replicates. The analytical data using this experimental scheme is shown in Table 3. The similarity of the six soil samples is then assessed by the statistical techniques referred to as *one-way analysis of variance.* Such a statistical analysis of the data is most easily performed using an ANOVA (ANalysis Of VAriance) table as illustrated in Table 4.

The total variation in the data can be partitioned between the variation amongst the sub-samples and the variation within the sub-samples. The computation proceeds by determining the sum of squares for each source of variation and then the variances.

The total variance for all replicates of all samples analysed is given, from Equation (3), by

$$s_T^2 = \sum_{j=1}^{m} \sum_{i=1}^{n} (x_{ij} - \bar{x})^2 / (N - 1) \tag{15}$$

where x_{ij} is the ith replicate of the jth sample. The total number of analyses is denoted by N, which is equal to the number of replicates per sample, n, multiplied by the number of samples, m. The numerator in Equation (15) is the sum of squares for the total variation, SS_T, and can be rearranged to simplify calculations,

$$SS_T = \sum_{j=1}^{m} \sum_{i=1}^{n} x_{ij}^2 - \left[\sum_{j=1}^{m} \sum_{i=1}^{n} x_{ij} \right]^2 / N \tag{16}$$

The variance among the different samples is obtained from SS_A,

$$SS_A = \sum_{j=1}^{m} \left[\sum_{i=1}^{n} x_{ij} \right]^2 / n - \left[\sum_{j=1}^{m} \sum_{i=1}^{n} x_{ij} \right]^2 / N \tag{17}$$

and the within-sample sum of squares, SS_W, can be obtained by difference,

$$SS_W = SS_T - SS_A \tag{18}$$

For the soil phosphate data, the completed ANOVA table is shown in Table 5.

Once the F-test value has been calculated it can be compared with standard tabulated values, using some pre-specified level of significance to check whether it lies in the critical region. If it does not, then there is no evidence to suggest that the samples arise from different sources and the hypothesis that all the values are similar can be accepted. From statistical tables, $F_{0.01, 5, 24} = 3.90$, and since the experimental value of 1.69 does not exceed this then the result is not significant at the 1% level and we can accept the hypothesis that there is no difference between the six sets of sub-samples.

The simple one-way analysis of variance discussed above can indicate the relative magnitude of differences in variance but provides no information as to

Table 5 *Completed ANOVA table for phosphate data from Table 3*

Source of variation	Sum of squares	Degrees of freedom	Mean squares	F-Test
Among samples	92.8	5	18.56	1.69
Within samples	264	24	11	
Total variation	356.8	29		

the source of the observed variation. For example, a single step in an experimental procedure may give rise to a large degree of error in the analysis. This would not be identified by ANOVA because it would be mixed with all other sources of variances in calculating SS_W. More sophisticated and elaborate statistical tests are readily available for a detailed analysis of such data and the interested reader is referred to the many statistics texts available.[3–8] The use of the F-test and analysis of variance will be encountered frequently in subsequent examples and discussions.

Outliers

The suspected presence of rogue values or outliers in a data set always causes problems for the analyst. Not only must we be able to detect them, but some systematic and reliable procedure for reducing their effect or eliminating them may need to be implemented. Methods for detecting outliers depend on the nature of the data as well as the data analysis being performed. For the present, two commonly employed methods will be discussed briefly.

The first method is Dixon's Q-test.[9] The data points are *ranked* and the difference between a suspected outlier and the observation closest to it is compared to the total range of measurements. This ratio is the Q-value. As with the t-test, if the computed Q-value is greater than tabulated critical values for some pre-selected level of significance, then the suspect data value can be identified as an outlier and may be rejected.

Use of this test can be illustrated with reference to the data in Table 6, which shows ten replicate measures of the molar absorptivity of nitrobenzene at 252 nm, its wavelength of maximum absorbance. Can the value of $\epsilon = 1056 \text{ mol}^{-1} \text{ m}^2$ be classed as an outlier? As defined above,

$$Q = |1056 - 1012|/|1056 - 990| = 0.67 \qquad (19)$$

For a sample size of 10, and with a 5% level of significance, the critical value of Q, from tables, is 0.464. The calculated Q-value exceeds this critical value, and therefore this point may be rejected from subsequent analysis. If necessary, the remaining data can be examined for further suspected outliers.

A second method involves the examination of *residuals*.[10] A residual is defined as the difference between an observed value and some expected, predicted or modelled value. If the suspect datum has a residual greater than, say, 4 times the residual standard deviation computed from all data, then it may be rejected. For the data in Table 6, the expected value is the mean of the ten results and the residuals are the differences between each value and this mean. The standard deviation of these residuals is 14.00 and the residual for the suspected outlier, 49, is certainly less than 4 times this value and, hence, this

[9] S.J. Haswell, in 'Practical Guide to Chemometrics', ed. S.J. Haswell, Marcel Dekker, New York, USA, 1992, p. 5.
[10] R.G. Brereton, 'Chemometrics', Ellis Horwood, Chichester, UK, 1990.

Table 6 *Molar absorptivity values for nitrobenzene measured at* 252 nm

e (mol^{-1} m^2 at 252 nm)				
1010	990	978	996	1005
1002	1056	1012	997	1004

point should not be rejected. If a 3σ criterion is employed then this datum is rejected.

If an outlier is rejected from a set of data then its value can be completely removed and the result discarded. Alternatively, the value can be replaced with an average value computed from all acceptable results or replaced by the next largest, or smallest, measure as appropriate.

Before leaving this brief discussion of outlier detection and treatment, a cautionary warning is appropriate. Testing for outliers should be strictly diagnostic, *i.e.* a means of checking that assumptions regarding the data distribution or some selected model are reasonable. Great care should be taken before rejecting any data; indeed there is a strong case for stating that no data should be rejected. If an outlier does exist, it may be more important to attempt to determine and address its cause, whether this be experimental error or some failure of the underlying model, rather than simply to reject it from the data. Outlier detection and treatment is of major concern to analysts, particularly with multivariate data where the presence of outliers may not be immediately obvious from visual inspection of tabulated data. Whatever mathematical treatment of outliers is adopted, visual inspection of graphical displays cf the data prior to and during analysis still remains one of the most effective means of identifying suspect data.

3 Lorentzian Distribution

Our discussions so far have been limited to assuming a normal, Gaussian distribution to describe the spread of observed data. Before proceeding to extend this analysis to multivariate measurements, it is worthwhile pointing out that other continuous distributions are important in spectroscopy. One distribution which is similar, but unrelated, to the Gaussian function is the *Lorentzian distribution*. Sometimes called the *Cauchy function*, the Lorentzian distribution is appropriate when describing resonance behaviour, and it is commonly encountered in emission and absorption spectroscopies. This distribution for a single variable, x, is defined by

$$f(x) = \frac{1}{\pi} \cdot \frac{\omega_{1/2}/2}{(x - \mu)^2 + (\omega_{1/2}/2)^2} \tag{20}$$

Like the normal distribution, the Lorentzian distribution is a continuous function, symmetric about its mean, μ, with a spread characterized by the

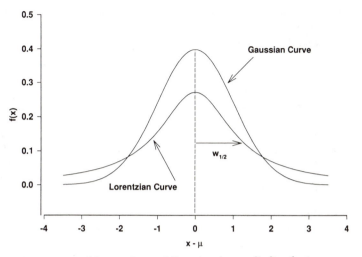

Figure 4 *Comparison of Lorentzian and Gaussian (normal) distributions*

half-width, $\omega_{1/2}$. The standard deviation is not defined for the Lorentzian distribution because of its slowly decreasing behaviour at large deviations from the mean. Instead, the spread is denoted by $\omega_{1/2}$, defined as the full-width at half maximum height. Figure 4 illustrates the comparison between the normal and Lorentzian shapes.[2] We shall meet the Lorentzian function again in subsequent chapters.

4 Multivariate Data

To this point, the data analysis procedures discussed have been concerned with a single measured variable. Although the determination of a single analyte constitutes an important part of analytical science, there is increasing emphasis being placed on multi-component analysis and using multiple measures in data analysis. The problems associated with manipulating and investigating multiple measurements on one or many samples constitutes that branch of applied statistics known as *multivariate analysis*, and this forms a major subject in chemometrics.[11–13]

Consideration of the results from a simple multi-element analysis will serve to illustrate terms and parameters associated with the techniques used. This example will also introduce some features of matrix operators basic to handling multivariate data.[14] In the scientific literature, matrix representation of multi-

[11] B.F.J. Manly, 'Multivariate Statistical Analysis: A Primer', Chapman and Hall, London, UK, 1991.
[12] A.A. Afifi and V. Clark, 'Computer Aided Multivariate Analysis', Lifetime Learning, California, USA, 1984.
[13] B. Flury and H. Riedwyl, 'Multivariate Statistics, A Practical Approach', Chapman and Hall, London, UK, 1988.
[14] M.J.R. Healy, 'Matrices for Statistics', Oxford University Press, Oxford, UK, 1986.

Table 7 *Results from the analysis of mineral water samples by atomic absorption spectrometry. Expressed as a data matrix, each column represents a variate and each row a sample or object*

Variables (mg kg^{-1})				
Samples	Sodium	Potassium	Calcium	Magnesium
1	10.8	1.6	41.3	7.2
2	7.1	1.1	72.0	8.0
3	14.1	2.0	92.0	8.2
4	17.0	3.1	117.0	18.0
5	5.7	0.4	47.5	16.5
6	11.3	1.8	62.2	14.6
Mean =	11.0	1.7	72.0	12.1
Variance =	17.8	0.8	812.8	23.3

variate statistics is common. For those readers unfamiliar with the basic matrix operations, or those who wish to refresh their memory, the Appendix provides a summary and overview of elementary and common matrix operations.

The data shown in Table 7 comprise a portion of a multi-element analysis of mineral water samples. The data from such an analysis can conveniently be arranged in an *n* by *m* array, where *n* is the number of objects, or samples, and *m* is the number of variables measured. This array is referred to as the *data matrix* and the purpose of using matrix notation is to allow us to handle arrays of data as single entities rather than having to specify each element in the array every time we perform an operation on the data set. Our data matrix can be denoted by the single symbol *X* and each element by x_{ij}, with the subscripts *i* and *j* indicating the number of the row and column respectively. A matrix with only one row is termed a row vector, *e.g.*, *r*, and with only one column, a column vector, *e.g.*, *c*.

Each measure of an analysed variable, or *variate*, may be considered independent. By summing elements of each column vector the mean and standard deviation for each variate can be calculated (Table 7). Although these operations reduce the size of the data set to a smaller set of descriptive statistics, much relevant information can be lost. When performing any multivariate data analysis it is important that the variates are not considered in isolation but are combined to provide as complete a description of the total system as possible. Interaction between variables can be as important as the individual mean values and the distributions of the individual variates. Variables which exhibit no interaction are said to be *statistically independent*, as a change in the value in one variable cannot be predicted by a change in another measured variable. In many cases in analytical science the variates are not statistically independent, and some measure of their interaction is required in order to interpret the data and characterize the samples. The degree or extent of this interaction between variables can be estimated by calculating their *covariances*, the subject of the next section.

Covariance and Correlation

Just as variance describes the spread of data about its mean value for a single variable, so the distribution of multivariate data can be assessed from the covariance. The procedure employed for the calculation of variance can be extended to multivariate analysis by computing the extent of the mutual variability of the variates about some common mean. The measure of this interaction is the covariance.

Equation (3), defining variance, can be written in the form,

$$s^2 = \sum x_{\mathrm{d}}^2/(n-1) \tag{21}$$

where $x_{\mathrm{d}} = x_i - \bar{x}$, or, in matrix notation,

$$s^2 = x_{\mathrm{d}}^{\mathrm{T}} \cdot x_{\mathrm{d}}/(n-1) \tag{22}$$

with x^{T} denoting the *transpose* of the column vector x to form a row vector (see Appendix). The numerator in Equations (21) and (22) is the corrected sum of squares of the data (corrected by subtracting the mean value and referred to as *mean centring*). To calculate covariance, the analogous quantity is the corrected sum of products, SP, which is defined by

$$SP_{jk} = \sum_{i=1}^{n} (x_{ij} - \bar{x}_j)(x_{ik} - \bar{x}_k) \tag{23}$$

where x_{ij} is the ith measure of variable j, *i.e.* the value of variable j for object i, x_{ik} is the ith measure of variable k, and SP_{jk} is the corrected sum of products between variables j and k. Note that in the special case where $j = k$ Equation (23) gives the sum of squares as used in Equation (3).

Sums of squares and products are basic to many statistical techniques and Equation (23) can be simply expressed, using the matrix form, as

$$SP = X_{\mathrm{d}}^{\mathrm{T}} \cdot X_{\mathrm{d}} \tag{24}$$

where X_{d} represents the data matrix after subtracting the column, *i.e.* variate, means. The calculation of variance is completed by dividing by $(n-1)$ and covariance is similarly obtained by dividing each element of the matrix SP by $(n-1)$.

The steps involved in the algebraic calculation of the covariance between sodium and potassium concentrations from Table 7 are shown in Table 8. The complete variance–covariance matrix for our data is given in Table 9.

For the data the variance–covariance matrix, COV_x, is *square*, the number of rows and number of columns are the same, and the matrix is *symmetric*. For a symmetric matrix, $x_{ij} = x_{ji}$, and some pairs of entries are duplicated. The covariance between, say, sodium and potassium is identical to that between potassium and sodium. The variance–covariance matrix is said to have *diagonal*

Table 8 *Calculation of covariance between sodium and potassium concentrations*

[Na] x_i	[K] x_j	$x_i - \bar{x}_i$	$x_j - \bar{x}_j$	$(x_i - \bar{x}_i)(x_j - \bar{x}_j)$
10.8	1.6	− 0.2	− 0.07	0.014
7.1	1.1	− 3.9	− 0.57	2.233
14.1	2.0	3.1	0.33	1.023
17.0	3.1	6.0	1.43	8.580
5.7	0.4	− 5.3	− 1.27	6.731
11.3	1.8	0.3	0.13	0.039
$\bar{x}$ = 11.0	1.67			18.610
Σ = 66.0	10.0			
s^2 = 17.81	0.82			

$$SP_{Na,K} = 128.61 - [(66.0)\cdot(10.0)]/6$$
$$= 18.61$$

$$COV_{Na,K} = 18.61/5 = 3.72$$

Table 9 *Symmetric variance–covariance matrix for the analytes in Table 7. The diagonal elements are the variances of individual variates; off-diagonal elements are covariances between variates*

	Sodium	Potassium	Calcium	Magnesium
Sodium	17.81	3.72	93.01	3.54
Potassium	3.72	0.82	20.59	0.91
Calcium	93.01	20.59	812.76	41.13
Magnesium	3.54	0.91	41.13	23.29

symmetry with the diagonal elements being the variances of the individual variates.

In Figure 5(a) a scatter plot of the concentration of sodium *vs.* the concentration of potassium, from Table 7, is illustrated. It can be clearly seen that the two variates have a high interdependence, compared with magnesium *vs.* potassium concentration, Figure 5(b). Just as the absolute value of variance is influenced by the units of measurement, so covariance is similarly affected. To estimate the degree of interrelation between variables, free from the effects of measurement units, the *correlation coefficient* can be employed. The linear correlation coefficient, r_{jk}, between two variables j and k is defined by,

$$r_{jk} = \text{Covariance}_{jk}/s_j \cdot s_k \qquad (25)$$

As the value for covariance can equal but never exceed the product of the standard deviations, values for r range from − 1 to + 1. The complete correlation matrix for the elemental data is presented in Table 10.

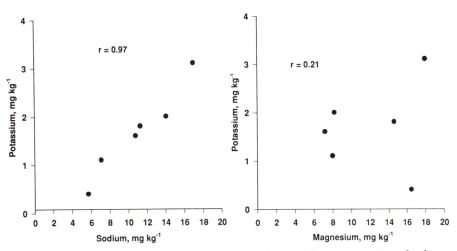

Figure 5   *There is a higher correlation, dependence, between the concentrations of sodium and potassium than between magnesium and potassium. Data from Table 7*

Figure 6 illustrates a series of scatter plots between variates having correlation coefficients between the two possible extremes. A correlation coefficient close to $+1$ indicates a high positive interdependence between variates, whereas a negative value means that the value of one variable decreases as the other increases, *i.e.* a strong negative interdependence. A value of r near zero indicates that the variables are linearly independent.

Correlation as a measure of similarity and association between variables is often used in many aspects of chemometrics. Used with care, it can assist in selecting variables for data analysis as well as providing a figure of merit as to how good a mathematical model fits experimental data, *e.g.* in constructing calibration curves. Returning to the extreme data set of Table 2, the correlation coefficient between chromium and nickel concentrations is identical for each source of water. If the data are plotted, however, some of the dangers of quoting r values are evident. From Figure 7, it is reasonable to propose a linear relationship between the concentrations of chromium and nickel for samples

Table 10   *Correlation matrix for the analytes in Table 7. The matrix is symmetric about the diagonal and values lie in the range -1 to $+1$*

	Sodium	Potassium	Calcium	Magnesium
Sodium	1.00	0.97	0.77	0.17
Potassium	0.97	1.00	0.80	0.21
Calcium	0.77	0.80	1.00	0.30
Magnesium	0.17	0.21	0.30	1.00

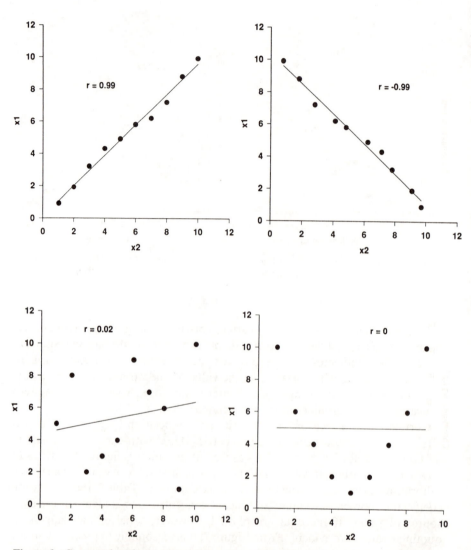

Figure 6 *Scatter plots for bivariate data with various values of correlation coefficient, r. Least-squares best-fit lines are also shown. Note that correlation is only a measure of linear dependence between variates*

from A. This is certainly not the case for samples B, and the graph suggests that a higher order, possibly quadratic, model would be better. For samples from source C, a potential outlier has reduced an otherwise excellent linear correlation, whereas for source D there is no evidence of any relationship between chromium and nickel but an outlier has given rise to a high correlation coefficient. To repeat the earlier warning, always visually examine the data before proceeding with any manipulation.

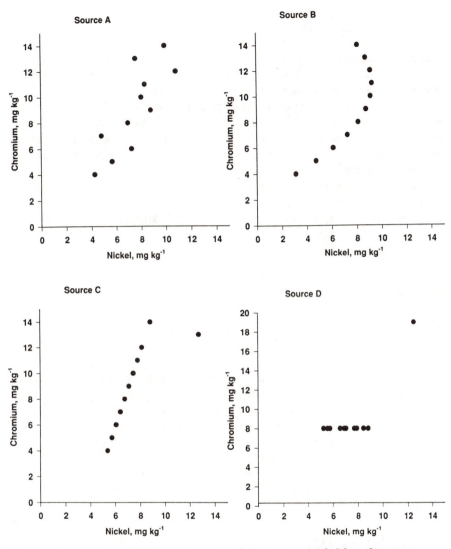

Figure 7 *Scatter plots of the concentrations of chromium vs. nickel from four waste water sources, from Table 2*

Multivariate Normal

In much the same way as the more common univariate statistics assume a normal distribution of the variable under study, so the most widely used multivariate models are based on the assumption of a multivariate normal distribution for each population sampled. The multivariate normal distribution is a generalization of its univariate counterpart and its equation in matrix notation is

$$f(x) = \frac{1}{(2\pi)^{m/2}|COV_x|^{1/2}} \exp[-\tfrac{1}{2}(x-\mu)^\mathrm{T}\, COV_x^{-1}(x-\mu)] \qquad (26)$$

The representation of this equation for anything greater than two variates is difficult to visualize, but the bivariate form ($m = 2$) serves to illustrate the general case. The exponential term in Equation (26) is of the form $x^\mathrm{T}Ax$ and is known as a *quadratic form* of a matrix product (Appendix A). Although the mathematical details associated with the quadratic form are not important for us here, one important property is that they have a well known geometric interpretation. All quadratic forms that occur in chemometrics and statistical data analysis expand to produce a quadratic surface that is a closed ellipse. Just as the univariate normal distribution appears bell-shaped, so the bivariate normal distribution is elliptical.

For two variates, x_1 and x_2, the mean vector and variance–covariance matrix are defined in the manner as discussed above.

$$x = \begin{bmatrix} x_1 \\ x_2 \end{bmatrix}, \quad \mu = \begin{bmatrix} \mu_1 \\ \mu_2 \end{bmatrix}$$

$$COV_x = \begin{bmatrix} \sigma_{11}^{\,2} & \sigma_{12}^{\,2} \\ \sigma_{21}^{\,2} & \sigma_{22}^{\,2} \end{bmatrix} \qquad (27)$$

where μ_1 and μ_2 are the means of x_1 and x_2 respectively, $\sigma_{11}^{\,2}$ and $\sigma_{22}^{\,2}$ are their variances, and $\sigma_{12}^{\,2} = \sigma_{21}^{\,2}$ is the covariance between x_1 and x_2. Figure 8 illustrates some bivariate normal distributions, and the contour plots show the lines of equal probability about the bivariate mean, *i.e.* lines that connect points having equal probability of occurring. The contour diagrams of Figure 8 may be compared to the correlation plots presented previously. As the covariance, $\sigma_{12}^{\,2}$, increases in a positive manner from zero, so the association between the variates increases and the spread is stretched, because the variables serve to act together. If the covariance is negative then the distribution moves in the other direction.

5 Displaying Data

As our discussions of population distributions and basic statistics have progressed, the use of graphical methods to display data can be seen to play an important role in both univariate and multivariate analysis. Suitable data plots can be used to display and describe both raw data, *i.e.* original measures, and transformed or manipulated data. Graphs can aid in data analysis and interpretation, and can serve to summarize final results.[15] The use of diagrams may help to reveal patterns in the data which may not be obvious from tabulated results. With most computer-based data analysis packages the graphics support

---

[15] J.M. Thompson, in 'Methods for Environmental Data Analysis', ed. C.N. Hewitt, Elsevier Applied Science, London, UK, 1992, p. 213.

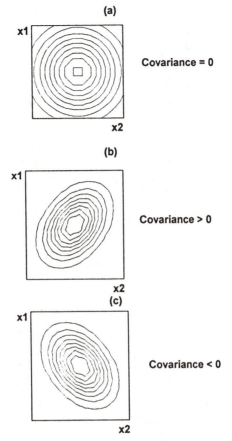

Figure 8 *Bivariate normal distributions as probability contour plots for data having different covariance relationships*

can provide a valuable interface between the user and the experimental data. The construction and use of graphical techniques to display univariate and bivariate data are well known. The common calibration graph or analytical working curve, relating, for example, measured absorbance to sample concentration, is ubiquitous in analytical science. No spectroscopist would welcome the sole use of tabulated spectra without some graphical display of the spectral pattern. The display of data obtained from more than two variables, however, is less common and a number of ingenious techniques and methods have been proposed and utilized to aid in the visualization of such multivariate data sets. With three variables a three-dimensional model of the data can be constructed and several graphical computer packages are available to assist in the design of three-dimensional plots.[15] In practice, the number of variables examined may well be in excess of two or three and less familiar and less direct techniques are required to display the data. Such techniques are generally referred to as *mapping methods* as they attempt to represent a many-

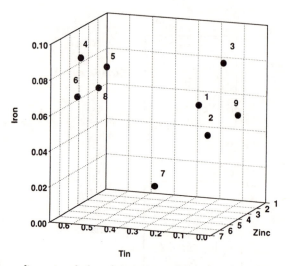

Figure 9   *Three dimensional plot of zinc, tin, and iron data from Table 11*

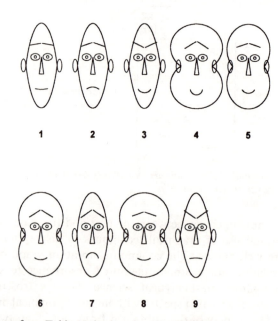

Figure 10   *Data from Table 11 displayed as Chernoff faces*

dimensional data set in a reduced, usually two-dimensional space whilst retaining the structure and as much information from the original data as possible.

For bivariate data the simple scatter plot of variate x against variate y is popular and there are several ways in which this can be extended to accommodate further variables. Figure 9 illustrates an example of a three-dimensional scatter plot. The data used are from Table 11, representing the results of the

Table 11 *XRF results from copper-based alloys*

Samples	Variables (% by weight)			
	Tin	Zinc	Iron	Nickel
1	0.20	3.40	0.06	0.08
2	0.20	2.40	0.04	0.06
3	0.15	2.00	0.08	0.16
4	0.61	6.00	0.09	0.02
5	0.57	4.20	0.08	0.06
6	0.58	4.82	0.07	0.02
7	0.30	5.60	0.02	0.01
8	0.60	6.60	0.07	0.06
9	0.10	1.60	0.05	0.19

analysis of nine alloys for four elements. The concentration of three analytes, zinc, tin, and iron, are displayed. It is immediately apparent from the illustration that the samples fall into one of two groups, with one sample lying between the groups. This pattern in the data is more readily seen in the graphical display than from the tabulated data.

This style of representation is limited to three variables and even then the diagrams can become confusing, particularly if there are a lot of points to plot. One method for graphically representing multivariate data ascribes each variable to some characteristic of a cartoon face. These *Chernoff faces* have been used extensively in the social sciences and adaptations have appeared in the analytical chemistry literature. Figure 10 illustrates the use of Chernoff faces to represent the data from Table 11. The size of the forehead is proportional to tin concentration, the lower face to zinc level, eyebrows to nickel, and mouth shape to iron concentration. As with the three-dimensional scatter plot, two groups can be seen, samples 1, 2, 3, and 9, and samples 4, 5, 6, and 8, with sample 7 displaying characteristics from both groups.

Star-plots present an alternative means of displaying the same data (Figure 11), with each ray size proportional to individual analyte concentrations.

A serious drawback with multidimensional representation is that visually some characteristics are perceived as being of greater importance than others and it is necessary to consider carefully the assignment of the variable to the graph structure. In scatter plots, the relationships between the horizontal co-ordinates can be more obvious than those for the higher-dimensional data on a vertical axis. It is usually the case, therefore, that as well as any strictly analytical reason for reducing the dimensionality of data, such simplification can aid in presenting multidimensional data sets. Thus, principal components and principal co-ordinates analysis are frequently encountered as graphical aids as well as for their importance in numerically extracting information from data. It is important to realize, however, that reduction of dimensionality can lead to loss of information. Two-dimensional representation of multivariate data can hide structure as well as aid in the identification of patterns.

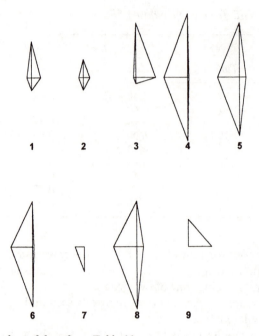

Figure 11   *Star plots of data from Table 11*

The wide variety of commercial computer software available to the analyst for statistical analysis of data has contributed significantly to the increasing use and popularity of multivariate analysis. It still remains essential, however, that the chemist appreciates the underlying theory and assumptions associated with the tests performed. In this chapter, only a brief introduction to the fundamental statistics has been presented. The remainder of the book is devoted to the acquisition, manipulation, and interpretation of spectrochemical data. No attempt has been made to present computer algorithms or program listings. Many fine texts are available that include details and listings of programs for numerical and statistical analysis for the interested reader.[16–19]

[16] A.F. Carley and P.H. Morgan, 'Computational Methods in the Chemical Sciences', Ellis Horwood, Chichester, UK, 1989.
[17] W.H. Press, B.P. Flannery, S.A. Teukolsky, and W.T. Vetterling, 'Numerical Recipes', Cambridge University Press, Cambridge, UK, 1987.
[18] J. Zupan, 'Algorithms for Chemists', Wiley, New York, USA, 1989.
[19] J.C. Davis, 'Statistics and Data Analysis in Geology', J. Wiley and Sons, New York, USA, 1973.

CHAPTER 2

Acquisition and Enhancement of Data

1 Introduction

In the modern spectrochemical laboratory, even the most basic of instruments is likely to be microprocessor controlled, with the signal output digitized. Given this situation, it is necessary for analysts to appreciate the basic concepts associated with computerized data acquisition and signal conversion to the digital domain. After all, digitization of the analytical signal may represent one of the first stages in the data acquisition and manipulation process. If this is incorrectly carried out then subsequent processing may not be worthwhile. The situation is analogous to that of analytical sampling. If a sample is not representative of the parent material, then no matter how good the chemistry or the analysis, the results may be meaningless or misleading.

The detectors and sensors commonly used in spectrometers are analogue devices; the signal output represents some physical parameter, *e.g.* light intensity, as a continuous function of time. In order to process such data in the computer, the continuous, or analogue, signal must be digitized to provide a series of numeric values equivalent to and representative of the original signal. An important parameter to be selected is how fast, or at what rate, the input signal should be digitized. One answer to the problem of selecting an appropriate sampling rate would be to digitize the signal at as high a rate as possible. With modern high-speed, analogue-to-digital converters, however, this would produce so much data that the storage capacity of the computer would soon be exceeded. Instead, it is preferred that the number of values recorded is limited. The analogue signal is digitally and discretely sampled, and the rate of sampling determines the accuracy of the digital representation as a time discrete function.

2 Sampling Theory

Figure 1 illustrates a data path in a typical ratio-recording, dispersive infrared spectrometer.[1] The digitization of the analogue signal produced by the detector

[1] M.A. Ford, in 'Computer Methods in UV, Visible and IR Spectroscopy', ed. W.O. George and H.A. Willis, The Royal Society of Chemistry, Cambridge, UK, 1990, p. 1.

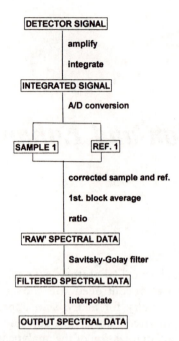

Figure 1 *Data path of a ratio-recording, dispersive IR spectrometer*
(Reproduced by permission from ref. 1)

is a critical step in the generation of the analytical spectrum. *Sampling theory* dictates that a continuous time signal can be completely recovered from its digital representation if the original analogue signal is *band-limited*, and if the sampling frequency employed for digitization is at least twice the highest frequency present in the analogue signal. This often quoted statement is fundamental to digitization and is worth examining in more detail.

The process of digital sampling can be represented by the scheme shown in Figure 2.[2] The continuous analytical signal as a function of time, x_t, is multiplied by a modulating signal comprising a train of pulses of equal magnitude and constant period, p_t. The resultant signal is a train of similar impulses but now with amplitudes limited by the spectral envelope x_t. We wish the digital representation accurately to reflect the original analogue signal in terms of all the frequencies present in the original data. Therefore, it is best if the signals are represented in the frequency domain (Figure 3). This is achieved by taking the Fourier transform of the spectrum.

Figure 3 illustrates the Fourier transform, x_f, of the analytical signal x_t.[2] At frequencies greater than some value, f_m, x_f is zero and the signal is said to be band-limited. Figure 3(b) shows the frequency spectrum of the modulating pulse train. The sampled signal, Figure 3(c), is repetitive with a frequency determined by the sampling frequency of the modulating impulses, f_s. These

[2] A.V. Openheim and A.S. Willsky, 'Signals and Systems', Préntice-Hall, New Jersey, USA, 1983.

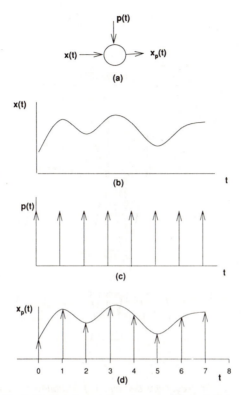

Figure 2 *A schematic of the digital sampling process: (a) A signal, x_t, is multiplied by a train of pulses, p_t, producing the signal $x_{p,t}$; (b) The analytical signal, x_t; (c) The carrier signal, p_t; (d) The resultant sampled signal is a train of pulses with amplitudes limited by x_t*
(Reproduced by permission from ref. 2)

modulating impulses have a period, t, given by $t = 1/f_s$. It is evident from Figure 3(c) that the sampling rate as dictated by the modulating signal, f_s, must be greater than the maximum frequency present in the spectrum, f_m. Not only that, it is necessary that the difference, $(f_s - f_m)$, must be greater than f_m, *i.e.*

$$(f_s - f_m) > f_m \quad \text{or} \quad f_s > 2f_m \tag{1}$$

$f_s = 2f_m$ is referred to as the minimum or *Nyquist sampling frequency*. If the sampling frequency, f_s, is less than the Nyquist value then *aliasing* arises. This effect is illustrated in Figure 3(d). At low sampling frequencies the spectral pattern is distorted by overlapping frequencies in the analytical data.

In practice, analytical signals are likely to contain a large number of very high-frequency components and, as pointed out above, it is impractical simply to go on increasing the digitizing rate. The situation may be relieved by applying a low pass filter to the raw analogue signal to remove high-frequency components and, hence, derive a band-limited analogue signal for subsequent

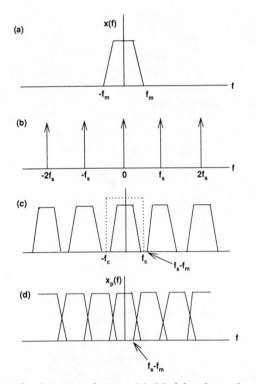

Figure 3 *Sampling in the frequency domain; (a) Modulated signal, x_t, has frequency spectrum x_f; (b) Harmonics of the carrier signal; (c) Spectrum of modulated signal is a repetitive pattern of x_f, and x_f can be completely recovered by low pass filtering using, for example, a box filter with cut-off frequency f_c; (d) Too low a sampling frequency produces aliasing, overlapping of frequency patterns*
(Reproduced by permission from ref. 2)

digital sampling. Provided that the high-frequency analogue information lost by this filtering is due only to noise, then the procedure is analytically valid. In the schematic diagram of Figure 1, this preprocessing function is undertaken by the integration stage prior to digitization.

Having digitized the analogue signal and obtained an accurate representation of the analytical information, the data can be manipulated further to aid the spectroscopist. One of the most common data processing procedures is digital filtering or smoothing to enhance the *signal-to-noise ratio*. Before discussing filtering, however, it will be worthwhile considering the concept of the signal-to-noise ratio and its statistical basis.

3 Signal-to-Noise Ratio

The spectral information used in an analysis is encoded as an electrical signal from the spectrometer. In addition to desirable analytical information, such signals contain an undesirable component termed *noise* which can interfere with the accurate extraction and interpretation of the required analytical data.

There are numerous sources of noise that arise from instrumentation, but briefly the noise will comprise flicker noise, interference noise, and white noise. These classes of noise signals are characterized by their frequency distribution. Flicker noise is characterized by a frequency power spectrum that is more pronounced at low frequencies than at high frequencies. This is minimized in instrumentation by modulating the carrier signal and using a.c. detection and a.c. signal processing, *e.g.* lock-in amplifiers. Interference from power supplies may also add noise to the signal. Such noise is usually confined to specific frequencies about 50 Hz, or 60 Hz, and their harmonics. By employing modulation frequencies well away from the power line frequency, interference noise can be reduced, and minimized further by using highly selective, narrow-bandpass electronic filters. White noise is more difficult to eliminate since it is random in nature, occurring at all frequencies in the spectrum. It is a fundamental characteristic of all electronic instruments. In recording a spectrum, complete freedom from noise is an ideal that can never be realized in practice. The noise associated with a recorded signal has a profound effect in an analysis and one figure of merit used to describe the quality of a measurement is the signal-to-noise ratio, S/N, which is defined as,

$$S/N = \frac{\text{average signal magnitude}}{\text{rms noise}} \tag{2}$$

The rms (room mean square) noise is the square root of the average deviation of the signal, x_i, from the mean noise value, *i.e.*

$$\text{rms noise} = \sqrt{\frac{\Sigma(\bar{x} - x_i)^2}{n - 1}} \tag{3}$$

This equation should be recognized as equating rms noise with the standard deviation of the noise signal, σ. S/N can, therefore, be defined as $\bar{x}/\sigma$.

In spectrometric analysis S/N is usually measured in one of two ways. The first technique is repeatedly to sample and measure the analytical signal and determine the mean and standard deviation using Equation (3). Where a chart recorder output is available, then a second method may be used. Assuming the noise is random and normally distributed about the mean, it is likely that 99% of the random deviations in the recorded signal will lie within $\mp 2.5\sigma$ of the mean value. By measuring the peak-to-peak deviation of the signal and dividing by 5, an estimate of the rms noise is obtained. The use of this method is illustrated in Figure 4. Whichever method is used, the signal should be sampled for sufficient time to allow a reliable estimate of the standard deviation to be made. When measuring S/N it is usually assumed that the noise is independent of signal magnitude for small signals close to the baseline or background signal.

Noise, as well as affecting the appearance of a spectrum, influences the sensitivity of an analytical technique and for quantitative analysis the S/N ratio

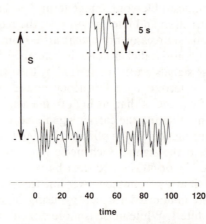

Figure 4 *Amplified trace of an analytical signal recorded with amplitude close to the background level, showing the mean signal amplitude, S, and the standard deviation, s. The peak-to-peak noise is 5s*

is of fundamental importance. Analytical terms dependent on the noise contained in the signal are the *decision limit*, the *detection limit*, and the *determination limit*. These analytical figures of merit are often quoted by instrument manufacturers and a knowledge of their calculation is important in evaluating and comparing instrument performance in terms of analytical sensitivity.

4 Detection Limits

The concept of an analytical detection limit implies that we can make a qualitative decision regarding the presence or absence of analyte in a sample. In arriving at such a decision there are two basic types of error that can arise (Table 1). The Type I error leads to the conclusion that the analyte is present in a sample when it is known not to be, and the Type II error is made if we conclude that the analyte is absent, when in fact it is present. The definition of a detection limit should address both types of error.[3]

Table 1 *The Type I and Type II errors that can be made in accepting or rejecting a statistical hypothesis*

	HYPOTHESIS IS CORRECT	HYPOTHESIS IS INCORRECT
HYPOTHESIS IS ACCEPTED	Correct decision	Type II Error
HYPOTHESIS IS REJECTED	Type I Error	Correct decision

[3] J.C. Miller and J.N. Miller, 'Statistics for Analytical Chemistry', Ellis Horwood, Chichester, UK, 1993.

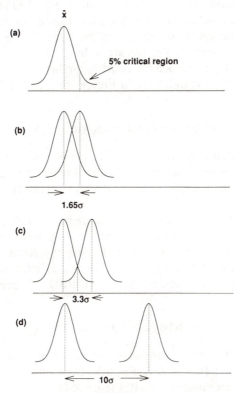

Figure 5 (a) *The normal distribution with the 5% critical region highlighted. Two normally distributed signals with equal variances overlapping, with the mean of one located at the 5% point of the other* (b) *– the decision limit; overlapping at their 5% points with means separated by 3.3σ* (c) *– the detection limit; and their means separated by 10σ* (d) *– the determination limit*

Consider an analytical signal produced by a suitable blank sample, with a mean value of μ_b. If we assume that noise in this background measurement is random and normally distributed about μ_b, then 95% of this noise will lie within $\mu_b \mp 1.65\sigma$ (Figure 5). With a 5% chance of committing a Type I error, then any analysis giving a response value greater than $\mu_b + 1.65\sigma$ can be assumed to indicate the presence of the analyte. This measure is referred to as the decision limit,

$$\text{Decision Limit} = z_{0.95} \cdot \sigma_b = 1.65\sigma_b \qquad (4)$$

If the number of measurements made to calculate σ_b is small, then the appropriate value from the *t*-distribution should be used in place of the *z*-value as obtained from the normal distribution curve.

We may ask, what if a sample containing analyte at a concentration equivalent to the decision limit is repeatedly analysed? In such a case, we can expect that in 50% of the measurements the analyte will be reported present, but in the other half of the measurements the analyte will be reported as not present. This attempt at defining a detection limit using the decision limit defined by Equation (4) does not address the occurrence of the Type II error.

If, as with the Type I error, we are willing to accept a 5% chance of committing a Type II error, then the relationship between the blank signal and sample measurement is as indicated in Figure 5(b). This defines the detection limit,

$$\text{Detection Limit} = 2z_{0.95}\sigma_b = 3.3\sigma_b \tag{5}$$

Under these conditions, we have a 5% chance of reporting the analyte present in a blank solution, and a 5% chance of reporting the analyte absent in a sample actually containing analyte at the concentration defined by the detection limit.

We should examine the precision of measurements made at this limit before accepting this definition of detection limit. The repeated measurement of the instrumental response from a sample containing analyte at the detection limit will lead to the analyte being reported as below the detection limit for 50% of the analyses. The relative standard deviation, RSD, of such measurements is given by

$$\text{RSD} = 100\sigma/\mu = 100/(2z_{0.95}) = 30.3\% \tag{6}$$

This hardly constitutes suitable precision for quantitative analysis, which should have a RSD of 10% or less. For a RSD of 10%, a further term can be defined called the determination limit, Figure 5(d),

$$\text{Determination Limit} = 10\sigma_b \tag{7}$$

When comparing methods, therefore, the defining equations should be identified and the definitions used should be agreed.

As we can see, the limits of quantitative analysis are influenced by the noise in the system and to improve the detection limit it is necessary to enhance the signal-to-noise ratio.

5 Reducing Noise

If we assume that the analytical conditions have been optimized, say to produce maximum signal intensity, then any increase in signal-to-noise ratio will be achieved by reducing the noise level. Various strategies are widely employed to reduce noise, including signal averaging, smoothing, and filtering. It is common in modern spectrometers for several methods to be used on the same analytical data at different stages in the data processing scheme (Figure 1).

Signal Averaging

The process of *signal averaging* is conducted by repetitively scanning and co-adding individual spectra. Assuming the noise is randomly distributed, then the analytical signals which are coherent in time are enhanced, since the signal grows linearly with the number of scans, N,

$$\text{signal magnitude} \propto N$$
$$\text{signal magnitude} = k_1 N \tag{8}$$

To consider the effect of signal averaging on the noise level we must refer to the *propagation of errors*. The variance associated with the sum of independent errors is equal to the sum of their variances, *i.e.*

$$\sigma_N^2 = \sum_{i=1}^{N} \sigma_i^2 = N\sigma_i^2 \tag{9}$$

Since we can equate rms noise with standard deviation then,

$$\sigma_N = \sqrt{(N\sigma_i^2)} \tag{10}$$

Thus the average magnitude of random noise increases at a rate proportional to the square root of the number of scans,

$$\text{noise magnitude} \propto N^{1/2}$$
$$\text{noise magnitude} = k_2 N^{1/2} \tag{11}$$

Therefore,

$$\frac{\text{signal}}{\text{noise}} = \frac{k_1 N}{k_2 N^{1/2}} = kN^{1/2} \tag{12}$$

and the signal-to-noise ratio is improved at a rate proportional to the square root of the number of scans. Figure 6 illustrates part of an infrared spectrum and the effect of signal averaging 4, 9, and 16 spectra. The increase in signal-to-noise ratio associated with increasing the number of co-added repetitive scans is evident.

For signal averaging to be effective, each scan must start at the same place in the spectrum otherwise analytical signals and useful information will also cancel and be removed. The technique is widely used but is most common in fast scanning spectrometers, particularly Fourier transform instruments such as NMR and IR. Co-adding one hundred scans is common in infrared spectroscopy in order to achieve a theoretical enhancement of 10:1 in signal-to-noise ratio. Whilst further gains can be achieved, practical considerations may limit the process. Even with a fast scan, say 1 s, the time required to perform 10 000 scans and aim to achieve a 100-fold improvement in signal-to-noise ratio may be unacceptable. In addition, computer memory constraints on storing the accumulated spectra may limit the maximum number of scans permitted.

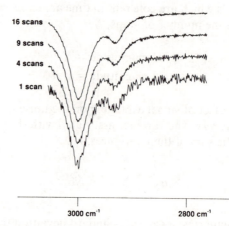

Figure 6   *An infrared spectrum and the results of co-adding 4, 9, and 16 scans from the same region*

Signal Smoothing

A wide variety of mathematical manipulation schemes are available to smooth spectral data, and in this section we shall concentrate on smoothing techniques that serve to average a section of the data. They are all simple to implement on personal computers. This ease of use has led to their widespread application, but their selection and tuning is somewhat empirical and depends on the application in-hand.

One simple smoothing procedure is *boxcar averaging*. Boxcar averaging proceeds by dividing the spectral data into a series of discrete, equally spaced, bands and replacing each band by a centroid average value. Figure 7 illustrates the results using the technique for different widths of the filter window or band. The greater the number of points averaged, the greater the degree of smoothing, but there is also a corresponding increase in distortion of the signal and subsequent loss of spectral resolution. The technique is derived from the use of electronic boxcar integrator units. It is less widely used in modern spectrometry than the methods of *moving average* and *polynomial smoothing*.

As with boxcar averaging, the moving average method replaces a group of values by their mean value. The difference in the techniques is that with the moving average successive bands overlap. Consider the spectrum illustrated in Figure 8, which is comprised of transmission values, denoted x_i. By averaging the first five values, $i = 1 \ldots 5$, a mean transmission value is produced which provides the value for the third data point, x'_3, in the smoothed spectrum. The procedure continues by incrementing i and averaging the next five values to find x'_4 from original data x_2, x_3, x_4, x_5, and x_6. The degree of smoothing achieved is controlled by the number of points averaged, *i.e.* the width of the smoothing window. Distortion of the data is usually less apparent with the moving average method than with boxcar averaging.

The mathematical process of implementing the moving average technique is

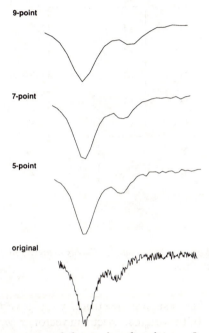

Figure 7 *An infrared spectrum and the results of applying a 5-point boxcar average, a 7-point average, and a 9-point average*

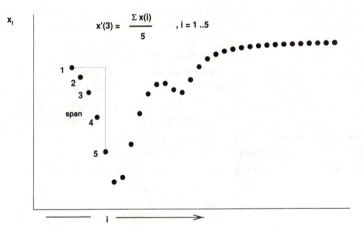

Figure 8 *Smoothing with a 5-point moving average. Each new point in the smoothed spectrum is formed by averaging a span of 5 points from the original data*

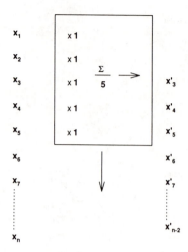

Figure 9   *Convolution of a spectrum with a filter is achieved by pulling the filter function across the spectrum*

termed *convolution*. The resultant spectrum, x' (as a vector), is said to be the result of convolution of the original spectrum vector, x, with a filter function, w, *i.e.*

$$x' = w \otimes x \tag{13}$$

For the simple five-point moving average, $w = [1,1,1,1,1]$. The mechanism and application of the convolution process can be visualized graphically as illustrated in Figure 9.

In 1964 Savitzky and Golay described a technique for smoothing spectral data using convolution filter vectors derived from the coefficients of least-squares-fit polynomial functions.[4] This paper, with subsequent arithmetic corrections,[5] has become a classic in analytical signal processing and least-squares polynomial smoothing is probably the technique in widest use in spectral data processing and manipulation. To appreciate its derivation and application we should extend our discussion of the moving average filter.

The simple moving average technique can be represented mathematically by

$$x'_i = \sum_{j=-n}^{n} x_{i+j}\omega_j \bigg/ \sum_{j=-n}^{n} \omega_j \tag{14}$$

where x_i and x'_i are elements of the original and smoothed data vectors respectively, and the values ω_j are the weighting factors in the smoothing window. For a simple moving average function, $\omega_j = 1$ for all j and the width of the smoothing function is defined by $(2n + 1)$ points.

[4] A. Savitzky and M.J.E. Golay, *Anal. Chem.*, 1964, **36**, 1627.
[5] J. Steiner, Y. Termonia, and J. Deltour, *Anal. Chem.*, 1972, **44**, 1906.

The process of polynomial smoothing extends the principle of the moving average by modifying the weight vector, ω, such that the elements of ω describe a convex polynomial. The central value in each window, therefore, adds more to the averaging process than values at the extremes of the window.

Consider five data points forming a part of a spectrum described by the data set x recorded at equal wavelength intervals. Polynomial smoothing seeks to replace the value of the point x_j by a value calculated from the least-squares polynomial fitted to x_{j-2}, x_{j-1}, x_j, x_{j+1}, and x_{j+2} recorded at wavelengths denoted by λ_{j-2}, λ_{j-1}, λ_j, λ_{j+1}, and λ_{j+2}.

For a quadratic curve fitted to the data, the model can be expressed as

$$x' = a_0 + a_1 \lambda + a_2 \lambda^2 \tag{15}$$

where x' is the fitted model data and a_0, a_1, and a_2 are the coefficients or weights to be determined.

Using the method of least squares, the aim is to minimize the error, ϵ, given by the square of the difference between the model function, Equation (13) and the observed data, for all data values fitted, *i.e.*

$$\epsilon = \Sigma(x_j' - x_j)^2 = \left[a_0 + a_1 \sum_{j=-n}^{n} \lambda_j + a_2 \sum \lambda_j^2 - \sum x_j \right]^2 \tag{16}$$

and, by simple differential calculus, this error function is a minimum when its first derivative is zero.

Differentiating Equation (16) with respect to a_0, a_1, and a_2 respectively, provides a set of so-called *normal equations*,

$$a_0 \sum_{j=-n}^{n} (1) + a_1 \sum \lambda_j + a_2 \sum \lambda_j^2 = \sum x_j$$

$$a_0 \sum \lambda_j + a_1 \sum \lambda_j^2 + a_2 \sum \lambda_j^3 = \sum x_j \lambda_j \tag{17}$$

$$a_0 \sum \lambda_j^2 + a_1 \sum \lambda_j^3 + a_2 \sum \lambda_j^4 = \sum x_j \lambda_j^2$$

Because the λ_j values are equally spaced, $\Delta\lambda = \lambda_j - \lambda_{j-1}$ is constant and only relative λ values are required for the model,

$$\lambda_j = j.\Delta\lambda \tag{18}$$

Hence, for $j = -2 \ldots +2$ (a five-point fit),

$$\begin{aligned}
\Sigma\lambda j^1 &= \Delta\lambda.\Sigma j^1 = 0 \\
\Sigma\lambda j^2 &= \Delta\lambda.\Sigma j^2 = 10\Delta\lambda \\
\Sigma\lambda j^3 &= \Delta\lambda.\Sigma j^3 = 0 \\
\Sigma\lambda j^4 &= \Delta\lambda.\Sigma j^4 = 34\Delta\lambda
\end{aligned} \tag{19}$$

which can be substituted into the normal equations, Equations (17), giving

$$5a_0 + 10\Delta\lambda^2.a_2 = \Sigma x_j = x_{j-2} + x_{j-1} + x_j + x_{j+1} + x_{j+2}$$
$$10\Delta\lambda.a_1 = \Sigma x_j.\lambda_j = -2x_{j-2} - x_{j-1} + x_{j+1} - 2x_{j+2}$$
$$10a_0 + 34\Delta\lambda^2.a_2 = \Sigma x_j.\lambda_j^2 = 4x_{j-2} + x_{j-1} + x_{j+1} + 4x_{j+2} \tag{20}$$

which can be rearranged,

$$a_0 = (-3x_{j-2} + 12x_{j-1} + 17x_j + 12x_{j+1} - 3x_{j+2}).1/35$$
$$a_1 = (-2x_{j-2} - x_{j-1} + x_{j+1} - 2x_{j+2}).1/10\Delta\lambda$$
$$a_2 = (2x_{j-2} - x_{j-1} - 2x_j - x_{j+1} + 2x_{j+2}).1/14\Delta\lambda^2 \tag{21}$$

At the central point in the smoothing window, $\lambda_j = 0$ and $x'_j = a_0$ from Equation (15). The five weighting coefficients, ω_j, are given by the first equation in Equation (21),

$$\omega = [-3, 12, 17, 12, -3] \tag{22}$$

Savitzky and Golay published the coefficients for a range of least-squares fit curves with up to 25-point wide smoothing windows for each.[4] Corrections to the original tables have been published by Steinier *et al.*[5]

Table 2 presents the weighting coefficients for performing 5, 9, 13, and 17-point quadratic smoothing and the results of applying these functions to the infrared spectral data are illustrated in Figure 10.

When choosing to perform a Savitzky–Golay smoothing operation on spectral data, it is necessary to select the filtering function (quadratic, quartic, *etc.*), the width of the smoothing function (the number of points in the smoothing window), and the number of times the filter is to be applied successively to the

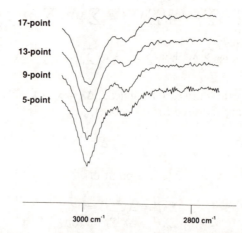

Figure 10   *Savitzky–Golay quadratic smoothing of the spectrum from Figure 7(a) using a 5-point plan (a), a 9-point span (b), a 13-point span (c), and a 17-point span (d)*

Table 2 *Savitzky–Golay coefficients, or weightings, for 5-, 9-, 13-, and 17-point quadratic smoothing of continuous spectral data*

Points	17	13	9	5
− 8	− 21			
− 7	− 6			
− 6	7	− 11		
− 5	18	0		
− 4	27	9	− 21	
− 3	34	16	14	
− 2	39	21	39	− 3
− 1	42	24	54	12
0	43	25	59	17
1	42	24	54	12
2	39	21	39	− 3
3	34	16	14	
4	27	9	− 21	
5	18	0		
6	7	− 11		
7	− 6			
8	− 21			
Norm	323	143	231	35

data. Although the final choice is largely empirical, the quadratic function is the most commonly used, with the window width selected according to the scanning conditions. A review and account of selecting a suitable procedure has been presented by Enke and Nieman.[6]

Filtering in the Frequency Domain

The smoothing operations discussed above have been presented in terms of the action of filters directly on the spectral data as recorded in the time domain. By converting the analytical spectrum to the frequency domain, the performance of these functions can be compared and a wide variety of other filters designed. Time-to-frequency conversion is accomplished using the Fourier transform. Its use was introduced earlier in this chapter in relation to sampling theory, and its application will be extended here.

The electrical output signal from a conventional scanning spectrometer usually takes the form of an amplitude–time response, *e.g.* absorbance *vs.* wavelength. All such signals, no matter how complex, may be represented as a sum of sine and cosine waves. The continuous function of composite frequencies is called a *Fourier integral*. The conversion of amplitude–time, t, information into amplitude-frequency, w, information is known as a Fourier transformation. The relation between the two forms is given by

6 C.G. Enke and T.A. Nieman, *Anal. Chem.*, 1976, **48**, 705.

$$F(w) = \int_{-\infty}^{\infty} f(t)[\cos(wt) + i.\sin(wt)]dt \qquad (23)$$

or, in complex exponential form,

$$F(w) = \int_{-\infty}^{\infty} f(t)e^{-2\pi iwt}\,dt \qquad (24)$$

The corresponding reverse, or inverse, transform, converting the complex frequency domain information back to the time domain is

$$f(t) = \int_{-\infty}^{\infty} F(w)e^{2\pi iwt}\,dw \qquad (25)$$

The two functions $f(t)$ and $F(w)$ are said to comprise *Fourier transform pairs*.

As discussed previously with regard to sampling theory, real analytical signals are band-limited. The Fourier equations therefore should be modified for practical use as we cannot sample an infinite number of data points. With this practical constraint, the discrete forward complex transform is given by

$$F(n) = \sum_{k=0}^{N-1} f(k)e^{-2\pi ikn/N} \qquad (26)$$

and the inverse is

$$f(k) = \frac{1}{N} \sum_{n=0}^{N-1} F(n)e^{2\pi ikn/N} \qquad (27)$$

A time domain spectrum consists of N points acquired at regular intervals and it is transformed to a frequency domain spectrum. This consists of $N/2$ real and $N/2$ imaginary data points, with $n = -N/2 \ldots 0 \ldots N/2$, and k takes integer values from 0 to $N-1$.

Once a frequency spectrum of a signal is computed then it can be modified mathematically to enhance the data in some well defined manner. The suitably processed spectrum can then be obtained by the inverse transform.

Several Fourier transform pairs are shown pictorially in Figure 11. An infinitely sharp amplitude–time signal, Figure 11(a), has a frequency response spectrum containing equal amplitudes at all frequencies. This is the white spectrum characteristic of a random noise amplitude–time signal. As the signal becomes broader, the frequency spectrum gets narrower. The higher frequencies are reduced dramatically and the frequency spectrum has the form $(\sin x)/x$, called the sinc function, Figure 11(b). For a triangular signal, Figure 11(c), the functional form of the frequency spectrum is $(\sin^2 x)/x^2$, the sinc^2 function. The sinc and sinc^2 forms are common filtering functions in interferometry, where their application is termed *apodisation*. The frequency response spectra of Lorentzian and Gaussian shaped signals are of particular

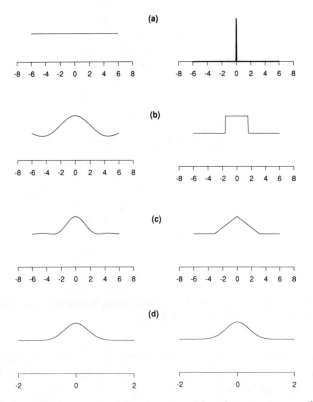

Figure 11 *Some well characterized Fourier pairs. The white spectrum and the impulse function* (a), *the boxcar and sinc functions* (b), *the triangular and sinc² functions* (c), *and the Gaussian pair* (d)
(Reproduced by permission from ref. 7)

interest since these shapes describe typical spectral profiles. The Fourier transform of a Gaussian signal is another Gaussian form, and for a Lorentzian signal the transform takes the shape of an exponentially decaying oscillator.

One of the earliest applications of the Fourier transform in spectroscopy was in filtering and noise reduction. This technique is still extensively employed.

Figure 12 presents the Fourier transform of an infrared spectrum, before and after applying the 13-point quadratic Savitzky–Golay function. The effect of smoothing can clearly be seen as reducing the high-frequency fluctuations, hopefully due to noise, by the polynomial function serving as a low-pass filter. Convolution provides an important technique for smoothing and processing spectral data, and can be undertaken in the frequency domain by simple multiplication. Thus smoothing can be accomplished in the frequency domain, following Fourier transformation of the data, by multiplying the Fourier transform by a rectangular or other truncating function. The low-frequency

[7] R. Bracewell, 'The Fourier Transform and Its Application', McGraw-Hill, New York, USA, 1965.

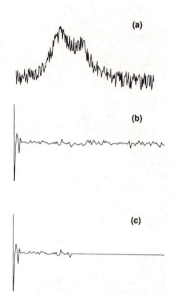

Figure 12 *A spectrum* (a) *and its Fourier transform before* (b) *and after applying a 13-point quadratic smoothing filter* (c)

Fourier coefficients should be relatively unaffected, whereas the high-frequency components characterizing random noise are reduced or zeroed. The subsequent inverse transform then yields the smoothed waveform.

The rectangular window function is a simple truncating function which can be applied to transformed data. This function has zero values above some pre-selected cut-off frequency, f_c, and unit values at lower frequencies. Using various cut-off frequencies for the truncating function and applying the inverse transform results in the smoothed spectra shown in Figure 13.

Although the selection of an appropriate cut-off frequency value is somewhat arbitrary, various methods of calculating a suitable value have been proposed in the literature. The method of Lam and Isenhour[8] is worth mentioning, not least because of the relative simplicity in calculating f_c. The process relies on determining what is termed the *equivalent width*, EW, of the narrowest peak in the spectrum. For a Lorentzian band the equivalent width in the time domain is given by

$$EW_t = \omega_{1/2} . \pi/2 \qquad (28)$$

where $\omega_{1/2}$ is the full-width at half-maximum, the half-width, of the narrowest peak in the spectrum.

The equivalent width in the frequency domain, EW_f, is simply the reciprocal

[8] R.B. Lam and T.L. Isenhour, *Anal. Chem.*, 1981, **53**, 1179.

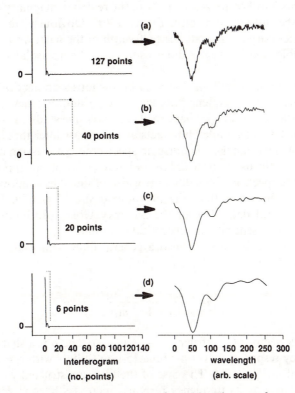

Figure 13 *A spectrum and its Fourier transform* (a). *The transform and its inverse retaining* (b) *40,* (c) *20, and* (d) *6 of the Fourier coefficients*

of EW_t. Assuming the spectrum was acquired in a single scan taking 10 s and it comprises 256 discrete points, then the sampling interval, Δt, is given by

$$\Delta t = 10/256 = 0.039 \text{ s} \tag{29}$$

and the maximum frequency, f_{max}, by

$$f_{max} = 1/(2\Delta t) = 12.75 \text{ Hz} \tag{30}$$

The IR spectrum was synthesized from two Lorentzian bands, the sharpest having $\omega_{1/2} = 1.17$ s. Therefore $EW_t = 1.838$ s and $EW_f = 0.554$ Hz.

The complex interferogram of 256 points is composed of 128 real values and 128 imaginary values spanning the range 0–12.75 Hz. According to the EW criterion, a suitable cut-off frequency is 0.554 Hz and the number of significant points, N, to be retained may be calculated from

$$N = (128)(0.554)/12.75 \cong 6 \tag{31}$$

Thus, points 7 to 128 are zeroed in both the real and imaginary arrays before performing the inverse transform, Figure 13(d). Obviously, to use the technique, it is necessary to estimate the half-width of the narrowest band present. Where possible this is usually done using some sharp isolated band in the spectrum.

All the smoothing functions discussed in previous sections can be displayed and compared in the frequency domain, and in addition new filters can be designed. Bromba and Ziegler have made an extensive study of such 'designer' filters.[9,10] The Savitzky–Golay filter acts as a low-pass filter that is optimal for polynomial shaped signals. Of course, in spectrometry Gaussian or Lorentzian band shapes are the usual form and the polynomial is only an approximation to a section of the spectrum defined by the width of the filter window. There is no reason why filters other than the polynomial should not be employed for smoothing spectral data. Use of the Savitzky–Golay procedure is as much traditional as representing any theoretical optimum.

Bromba and Ziegler have defined a general filter with weighting elements defined by the form

$$\omega_j = \frac{2\alpha + 1}{2n + 1} - \frac{\alpha|j|}{n(n + 1)} \tag{32}$$

where ω is the vector of coefficients, $j = -n \ldots n$, and α is a shape parameter. The frequency-response curves of three 15-point filters with $\alpha = 0.5$, 1, and 2 are illustrated in Figure 14. The case of the filter constructed with $\alpha = 2$ is of particular interest as its frequency response increases beyond zero frequency and then falls off rapidly. The effect of convoluting a spectrum with this function is apparently to enhance resolution. The practical use of such filters should be undertaken with care, however, and they are best used in an interactive mode when the user can visibly assess the effects before proceeding to further data manipulation.

Whatever smoothing technique is employed, the aim is to reduce the effects of random variations superimposed on the analytically useful signal. This transform can be simply expressed as

$$\text{Spectrum (smoothed)} = \text{Spectrum (raw)} - \text{noise} \tag{33}$$

Assuming all noise is removed then the result is the true spectrum. Conversely, from Equation (33), if the smoothed spectrum is subtracted from the original, raw data, then a noise spectrum is obtained. The distribution of this noise as a function of wavelength may provide information regarding the source of the noise in spectrometers. The procedure is analogous to the analysis of residuals in regression analysis and modelling.

[9] M.U.A. Bromba and H. Ziegler, *Anal. Chem.*, 1983, **55**, 1299.
[10] M.U.A. Bromba and H. Ziegler, *Anal. Chem.*, 1983, **55**, 648.

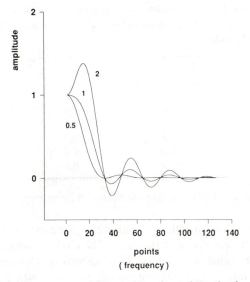

Figure 14 *The frequency response of filters of Bomba and Ziegler for a values of 0.5, 1.0, and 2.0*

6 Interpolation

Not all analytical data can be recorded on a continuous basis; discrete measurements often have to be made and they may not be at regular time or space intervals. To predict intermediate values for a smooth graphic display, or to perform many mathematical manipulations, *e.g.* Savitzky–Golay smoothing, it is necessary to evaluate regularly spaced intermediate values. Such values are obtained by *interpolation*.

Obviously, if the true underlying mathematical relationship between the independent and dependent variables is known then any value can be computed exactly. Unfortunately, this information is rarely available and any required interpolated data must be estimated.

The data in Table 3, shown in Figure 15, consist of magnesium concentrations as determined from river water samples collected at various distances from the stream mouth. Because of the problems of accessibility to sampling sites, the samples were collected at irregular intervals along the stream channel and the distances between samples were calculated from aerial photographs. To produce regularly spaced data, all methods for interpolation assume that no discontinuity exists in the recorded data. It is also assumed that any intermediate, estimated value is dependent on neighbouring recorded values. The simplest interpolation technique is *linear interpolation*. With reference to Figure 15, if y_1 and y_2 are observed values at points x_1 and x_2, then the value of y' situated at x' between x_1 and x_2 can be calculated from

$$y' = y_1 + \frac{(y_2 - y_1)(x' - x_1)}{(x_2 - x_1)} \tag{34}$$

Table 3 *Concentration of maganesium* (mg kg^{-1}) *from a stream sampled at different locations along its course. Distances are from stream mouth to sample locations*

Distance (m)	Mg (mg kg^{-1})
1800	4.0
2700	10.1
4500	11.5
5200	10.2
7100	8.4
8500	8.6

For a value of x' of 2500 m the estimated magnesium concentration, y', is 8.74 mg kg^{-1}.

The difference between values of adjacent points is assumed to be linear function of the distance separating them. The closer a point is to an observation, the closer its value is to that of the observation. Despite the simplicity of the calculation, linear interpolation should be used with care as the abrupt changes in slope that may occur at recorded values are unlikely to reflect accurately the more smooth transitions likely to be observed in practice. A better, and graphically more acceptable, result is achieved by fitting a smooth curve to the data. Suitable polynomials offer an excellent choice.

Polynomial interpolation is simply an extension of the linear method. The polynomial is formed by adding extra terms to the model to represent curved regions of the spectrum and using extra data values in the model.

If only one pair of measurements had been made, say (y_1, x_1), then a zeroth order equation of the type $y' = y_1$, for all y' would be the only possible solution. With two pairs of measurements, (y_1, x_1) and (y_2, x_2), then a first-order linear model can be proposed,

$$y' = y_1 + a_0(x' - x_1) \tag{35}$$

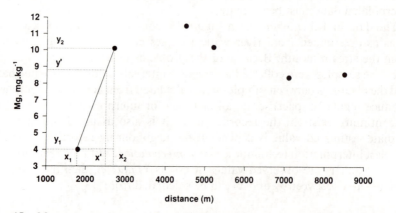

Figure 15 *Magnesium concentration as a function of distance from the stream source and the application of linear interpolation*

where

$$a_0 = (y_2 - y_1)/(x_2 - x_1) \tag{36}$$
$$= 6.77 \times 10^{-3} \text{ for the magnesium data.}$$

This, of course, is the model of linear interpolation and for $x' = 2500$ m, $y' = 8.74$ mg kg^{-1}.

To take account of more measured data, higher order polynomials can be employed. A quadratic model will fit three pairs of points,

$$y' = y_1 + a_0(x' - x_1) + a_1(x' - x_1)(x' - x_2) \tag{37}$$

with the quadratic term being zero when $x' = x_1$ or $x' = x_2$. When $x' = x_3$ then substitution and rearrangement of Equation (37) allows the coefficient a_1 to be calculated,

$$y_3 = y_1 + \frac{(y_2 - y_1)(x_3 - x_1)}{(x_2 - x_1)} + a_1(x_3 - x_1)(x_3 - x_2) \tag{38}$$

and

$$a_1 = \frac{\dfrac{(y_3 - y_1)}{(x_3 - x_1)} - \dfrac{(y_2 - y_1)}{(x_2 - x_1)}}{(x_3 - x_2)} \tag{39}$$

$$= -2.2 \times 10^{-6}, \text{ for the magnesium data.}$$

Substituting for a_1 and $x' = 2500$ m into Equation (37), the estimated value of y' is 9.05 mg kg^{-1} Mg.

The technique can be extended further. With four pairs of observations, a cubic equation can be generated to pass through each point,

$$y' = y_1 + a_0(x' - x_1) + a_1(x' - x_1)(x' - x_2) + a_2(x' - x_1)(x' - x_2)(x' - x_3) \tag{40}$$

and by a similar process, at x_4 the coefficient a_2 is given by

$$a_2 = \frac{\dfrac{\dfrac{(y_4 - y_1)}{(x_4 - x_1)} - \dfrac{(y_2 - y_1)}{(x_2 - x_1)}}{(x_4 - x_2)} - \dfrac{\dfrac{(y_3 - y_1)}{(x_3 - x_1)} - \dfrac{(y_2 - y_1)}{(x_2 - x_1)}}{(x_3 - x_2)}}{(x_4 - x_3)} \tag{41}$$

$$= -4.28 \times 10^{-10}, \text{ for the magnesium data.}$$

and substituting into Equation (40), for $x' = 2500$ m, then $y' = 8.93$ mg kg^{-1} Mg.

As the number of observed points to be connected increases, then so too does the degree of the polynomial required if we are to guarantee passing through each point. The general technique is referred to as providing *divided difference* polynomials. The coefficients a_2, a_3, a_4, *etc.* may be generated algorithmically by the '*Newton forward formula*', and many examples of the algorithms are available.[11,12]

To fit a curve to *n* data points a polynomial of degree $(n - 1)$ is required, and with a large data set the number of coefficients to be calculated is correspondingly large. Thus 100 data points could be interpolated using a 99-degree polynomial. Polynomials of such a high degree, however, are unstable. They can fluctuate wildly with the high-degree terms forcing an exact fit to the data. Low-degree polynomials are much easier to work with analytically and they are widely used for curve fitting, modelling, and producing graphic output. To fit small polynomials to an extensive set of data it is necessary to abandon the idea of trying to force a single polynomial through all the points. Instead different polynomials are used to connect different segments of points, piecing each section smoothly together. One technique exploiting this principle is *spline interpolation*, and its use is analogous to using a mechanical flexicurve to draw manually a smooth curve through fixed points.

The shape described by a spline between two adjacent points, or *knots*, is a cubic, third-degree polynomial. For the six pairs of data points representing our magnesium study, we would consider the curve connecting the data to comprise five cubic polynomials. Each of these take the form

$$s_i(x) = a_i x^3 + b_i x^2 + c_i x + d_i, \qquad i = 1 \ldots 5 \tag{42}$$

To compute the spline, we must calculate values for the 20 coefficients, four for each polynomial segment. Therefore we require 20 simultaneous equations, dictated by the following physical constraints imposed on the curve.

Since the curve must touch each point then

$$\begin{aligned} s_i(x_i) &= y_i, & i &= 1 \ldots 5 \\ s_i(x_{i+1}) &= y_{i+1}, & i &= 1 \ldots 5 \end{aligned} \tag{43}$$

The spline must curve smoothly about each point with no sharp bends or kinks, so the slope of each segment where they connect must be similar. To achieve this the first derivatives of the spline polynomials must be equal at the measured points.

$$\frac{\mathrm{d}s_{i-1}(x_i)}{\mathrm{d}x} = \frac{\mathrm{d}s_i(x_i)}{\mathrm{d}x}, \qquad i = 2 \ldots 5 \tag{44}$$

We can also demand that the second derivatives of each segment will be similar at the knots.

[11] A.F. Carley and P.H. Morgan, 'Computational Methods in the Chemical Sciences', Ellis Horwood, Chichester, UK, 1989.
[12] P. Gans, 'Data Fitting in the Chemical Sciences', J. Wiley and Sons, Chichester, UK, 1992.

$$\frac{d^2 s_{i-1}(x_i)}{dx^2} = \frac{d^2 s_i(x_i)}{dx^2}, \quad i = 2 \ldots 5 \tag{45}$$

Finally, we can specify that at the extreme ends of the curve the second derivatives are zero:

$$\frac{d^2 s_1(x_1)}{dx^2} = 0$$

$$\frac{d^2 s_5(x_6)}{dx^2} = 0 \tag{46}$$

From Equations (43) to (46) we can derive our 20 simultaneous equations and, by suitable rearrangement and substitution of values for x and y, determine the values of the 20 coefficients a_i, b_i, c_i, and d_i, $i = 1 \ldots 5$.

This calculation is obviously laborious and the same spline can be computed more efficiently by suitable scaling and substitution in the equations.[11] If the value of the second derivative of the spline at x_i is represented by p_i,

$$p_i = \frac{d^2 s_{i-1}(x_i)}{dx^2} = \frac{d^2 s_i(x_i)}{dx^2}, \quad i = 2 \ldots 5 \tag{47}$$

then if the values of $p_1 \ldots p_5$ were known, all the coefficients, a, b, c, d, could be computed from the following four equations,

$$s_i(x_i) = y_i$$
$$s_i(x_{i+1}) = y_{i+1}$$
$$\frac{d^2 s_i(x_i)}{dx^2} = p_i$$
$$\frac{d^2 s_i(x_{i+1})}{dx^2} = p_{i+1} \tag{48}$$

If each spline segment is scaled on the x-axis between the limits $[0,1]$, using the term $t = (x - x_i)/(x_{i+1} - x_i)$, then the curve can be expressed as[11]

$$s_i(t) = \frac{t y_{i+1} + (1 - t) y_i + (x_{i+1} - x_i)^2 \{(t^3 - t) p_{i+1} - [(1 - t)^3 - (1 - t)] p_i\}}{6} \tag{49}$$

To calculate the values of p_i, we impose the constraint that the first derivatives of the spline segments are equal at their endpoints. The resulting equations are

$$v_2 p_2 + u_2 p_3 = w_2$$
$$u_2 p_2 + v_3 p_3 + u_3 p_4 = w_3$$
$$u_3 p_3 + v_4 p_4 + u_4 p_5 = w_4$$
$$u_4 p_5 + v_5 p_5 = w_5 \tag{50}$$

52

or in matrix form,

$$
\begin{bmatrix}
v_2 & u_2 & 0 & 0 \\
u_2 & v_3 & u_3 & 0 \\
0 & u_3 & v_4 & u_4 \\
0 & 0 & u_4 & v_5
\end{bmatrix}
\cdot
\begin{bmatrix} p_2 \\ p_3 \\ p_4 \\ p_5 \end{bmatrix}
=
\begin{bmatrix} w_2 \\ w_3 \\ w_4 \\ w_5 \end{bmatrix}
\tag{51}
$$

where

$$u_i = x_{i+1} - x_i, \qquad v_i = 2(x_{i+1} - x_{i-1})$$

$$w_i = 6\left(\frac{(y_{i+1} - y_i)}{(x_{i+1} - x_i)} - \frac{(y_i - y_{i-1})}{(x_i - x_{i-1})}\right) \tag{52}$$

Equation (51) can be solved for p_i by conventional elimination methods.

Once the p_i values have been computed, the value of t for any segment can be calculated. From this the spline, $s_i(x)$, can be determined using Equation (49) and the appropriate values for p_i and p_{i+1}.

For the magnesium in river water data, after scaling the distance data to km, we have,

$$
\begin{bmatrix}
5.4 & 1.8 & 0.0 & 0.0 \\
1.8 & 5.0 & 0.7 & 0.0 \\
0.0 & 0.7 & 5.2 & 1.9 \\
0.0 & 0.0 & 1.9 & 6.6
\end{bmatrix}
\cdot
\begin{bmatrix} p_2 \\ p_3 \\ p_4 \\ p_5 \end{bmatrix}
=
\begin{bmatrix} -36 \\ -16 \\ 5.5 \\ 6.5 \end{bmatrix}
\tag{53}
$$

with the result $p_2 = -6.314$, $p_3 = -1.058$, $p_4 = 0.939$, and $p_5 = 0.714$.

To estimate the magnesium concentration at a distance $x = 2.5$ km, a value of t is calculated,

$$t = (x' - x_1)/(x_2 - x_1) = 0.778 \tag{54}$$

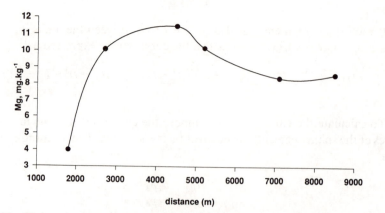

Figure 16 *The result of applying a cubic spline interpolation model to the stream magnesium data*

and this, with values for $p_1 = 0$ and $p_2 = -6.314$, is substituted into Equation (49),

$$s_1(t) = ty_2 + (1 - t)y_1 + (x_2 - x_1)^2(t^3 - t)p_2/6 \qquad (55)$$
$$s_1(t) = 9.008 \text{ mg kg}^{-1} \text{ Mg}$$

The resultant cubic spline curve for the complete range of the magnesium data is illustrated in Figure 16.

Spline curve fitting has many important applications in analytical science, not only in interpolation but also in differentiation and calibration. The technique is particularly useful when no analytical model of the data is available.[12]

Having acquired our chemical data, it is now necessary to analyse the results and extract the required relevant information. This will obviously depend on the aims of the analysis, but further preprocessing and manipulation of the data may be needed. This is considered in the next chapter.

CHAPTER 3

Feature Selection and Extraction

1 Introduction

Previous chapters have largely been concerned with processes related to acquiring our analytical data in a digital form suitable for further manipulation and analysis. This data analysis may include calibration, modelling, and pattern recognition. Many of these procedures are based on multivariate numerical data processing and before the methods can be successfully applied it is usual to perform some pre-processing on the data. There are three main aims of this pre-processing stage in data analysis,

(a) to reduce the amount of data and eliminate data that are irrelevant to the study being undertaken,
(b) to preserve or enhance sufficient information within the data in order to achieve the desired goal,
(c) to extract the information in, or transform the data to, a form suitable for further analysis.

One of the most common forms of pre-processing spectral data is *normalization*. At its simplest this may involve no more than scaling each spectrum in a collection so that the most intense band in each spectrum is some constant value. Alternatively, spectra could be normalized to constant area under the curve of the absorption or emission profile. A more sophisticated procedure involves constructing a covariance matrix between variates and extracting the *eigenvectors* and *eigenvalues*. Eigen analysis yields a set of new variables which are linear combinations of the original variables. This can often lead to representing the original information in fewer new variables, thus reducing the dimensionality of the data and aiding subsequent analysis.

The success of pattern recognition techniques can frequently be enhanced or simplified by suitable prior treatment of the analytical data, and *feature selection* and *feature extraction* are important stages in chemometrics. Feature selection refers to identifying and selecting those features present in the analytical data which are believed to be important to the success of calibration or pattern recognition. Techniques commonly used include *differentiation, integration*, and *peak identification*. Feature extraction, on the other hand, changes the dimensionality of the data and generally refers to processes combining or transforming original variables to provide new and better variables. Methods widely used include Fourier transformation and *principal components analysis.* In this chapter the popular techniques pertinent to feature selection and extraction are introduced and developed. Their application is illustrated with reference to spectrochemical analysis.

2 Differentiation

Derivative spectroscopy provides a means for presenting spectral data in a potentially more useful form than the zero'th order, normal data. The technique has been used for many years in many branches of analytical spectroscopy. Derivative spectra are usually obtained by differentiating the recorded signal with respect to wavelength as the spectrum is scanned. Whereas early applications mainly relied on hard-wired units for electronic differentiation, modern derivative spectroscopy is normally accomplished computationally using mathematical functions. First-, second-, and higher-order derivatives can easily be generated.

Analytical applications of derivative spectroscopy are numerous and generally owe their popularity to the apparent higher resolution of the differential

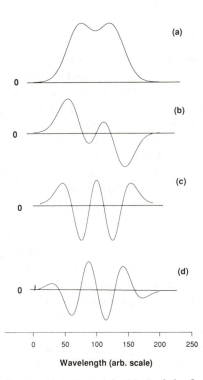

Figure 1 *A pair of overlapping Gaussian peaks* (a), *and the first-* (b), *second-* (c), *and third-order* (d) *derivative spectra*

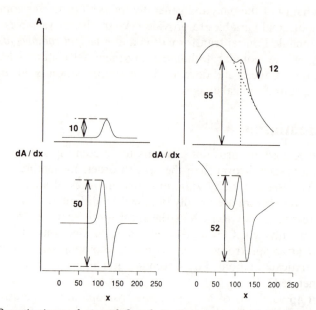

Figure 2 *Quantitative analysis with first derivative spectra. Peak heights are displayed in relative absorbance units*

data compared with the original spectrum. The effect can be illustrated with reference to the example shown in Figure 1. The zero'th-, first-, and second-order derivatives of a spectrum, comprised of the sum of two overlapping Gaussian peaks, are presented. The presence of a smaller analyte peak can be much more evident in the derivative spectra. In addition, for determining the intensity of the smaller peak in the presence of the large neighbouring peak, derivative spectra can be more useful and may be subject to less error. This is illustrated in Figure 2, in which the zero'th and first derivative spectra are shown for an analyte band with and without the presence of an overlapping band. If we assign unit peak height to the analyte in the normal, zero'th-order, spectrum, then for the same band with the interfering band present, a peak height of 55 units is recorded. Using a tangent baseline in order to attempt to correct for the overlap fails as there is no unique or easily identified tangent, and a not unreasonable value of 12 units for the peak height could be recorded, a 20% error. The situation is improved considerably if the first derivative spectrum is analysed. A value of one is assigned to the peak-to-peak distance of the lone analyte spectrum. In the presence of the overlapping band a similar measure for the analyte is now 1.04, a 4% error.

This example, however, oversimplifies the case of using derivative spectroscopy as it gives no indication of the effects of noise on the results. Derivative spectra tend to emphasize changes in slope that are difficult to detect in the zero'th-order spectrum. Unfortunately, as we have seen in previous chapters, noise is often comprised of high-frequency components and thus may be greatly amplified by differentiation. It is the presence of noise which generally limits the use of derivative spectroscopy to UV–visible spectrometry and other techniques in which a high signal-to-noise ratio may be obtained for a spectrum.

Various mathematical procedures may be employed to differentiate spectral data. We will assume that such data are recorded at evenly spaced intervals along the wavelength, λ, or other x-axis. If this is not the case, the data may be interpolated to provide this. The simplest method to produce the first-derivative spectrum is by difference,

$$\frac{dy}{d\lambda} = \frac{y_{i+1} - y_i}{\Delta\lambda} \tag{1}$$

or,

$$\frac{dy}{d\lambda} = \frac{y_{i+1} - y_{i-1}}{2\Delta\lambda} \tag{2}$$

and for the second derivative,

$$\frac{d^2y}{d\lambda^2} = \frac{y_{i+1} - 2y_0 - y_{i-1}}{\Delta\lambda^2} \tag{3}$$

where y represents the spectral intensity, the absorbance, or other metric.

Various other methods have been proposed to compute derivatives, including the use of suitable polynomial derivatives as suggested by Savitzky and Golay.[1,2] The use of a suitable array of weighting coefficients as a smoothing function with which to convolute spectral data was described in Chapter 2. In a similar manner, an array can be specified which on convolution produces the first, or higher degree differential spectrum. Using a quadratic polynomial and a five-point moving window, the first derivative is given by

$$\frac{dy}{d\lambda} = \frac{1}{10\Delta\lambda}(-2y_{i-2} - y_{i-1} + y_{i+1} + 2y_{i+2}) \tag{4}$$

and for the second derivative,

$$\frac{d^2y}{d\lambda^2} = \frac{1}{7\Delta\lambda^2}(2y_{i-2} - y_{i-1} - 2y_i - y_{i+1} + 2y_{i+2}) \tag{5}$$

Equations (4) and (5) are similar to Equations (2) and (3). The difference is in the use of additional terms using extra points from the data in order to provide a better approximation.

The relative merits of these different methods can be compared by differen-

[1] A. Savitzky and M.J.E. Golay, *Anal. Chem.*, 1964, **36**, 1627.
[2] P. Gans, 'Data Fitting in the Chemical Sciences', J. Wiley and Sons, Chichester, UK, 1992.

Table 1   *Derivative model data, $y = x + x^2/2 + noise$*

x	0	1	2	3	4	
Noise	0.200	0.000	0.200	− 0.200	0.200	
y	0.000	1.500	4.000	7.500	12.000	
y + Noise	0.200	1.500	4.200	7.300	12.200	Data 1
y + (Noise/2)	0.100	1.500	4.100	7.400	12.100	Data 2
y + (Noise/4)	0.050	1.500	4.050	7.450	12.050	Data 3
y + (Noise/8)	0.025	1.500	4.025	7.475	12.025	Data 4

Table 2   *Derivatives of $y = x + x^2/2 + noise$ (from Table 1) determined by difference formulae*

		Data			
		1	2	3	4
$dy/d\lambda$	by Equation (1)	3.1 or 2.7	3.3 or 2.6	3.4 or 2.55	3.45 or 2.525
$dy/d\lambda$	by Equation (2)	2.9	2.95	2.98	2.9875
$dy/d\lambda$	by Equation (4)	2.980	2.990	2.995	2.9975
$d^2y/d\lambda^2$	by Equation (3)	0.4	0.7	0.85	0.925
$d^2y/d\lambda^2$	by Equation (5)	1.0857	1.0429	1.0214	1.0107

tiating a known mathematical function. The model we will use is $y = (x + x^2/2)$ at the point $x = 2$. Various levels of noise are imposed on the signal y, as shown in Table 1. The resulting derivatives are shown in Table 2. As the noise level reduces and tends to zero, the derivative results from applying the five-point polynomial converge more quickly towards the correct noise-free value of 3 for the first derivative, and 1 for the second derivative. As with polynomial smoothing, the Savitzky–Golay differentiation technique is available with many commercial spectrometers.

Just as smoothing can be undertaken in the frequency domain (as discussed in Chapter 2), so too can differentiation following Fourier transformation of the amplitude–time spectrum. Obtaining a derivative in the Fourier domain is quite simple and is achieved by multiplying the transform by a linear function.[3] The effect is illustrated in Figure 3. The original spectrum, comprising two overlapping bands and random noise, is first converted by Fourier transformation to the frequency domain. The transformed data are multiplied by the filter function shown in Figure 3(c), and the result transformed back to produce the first-derivative spectrum, Figure 3(d). The susceptibility of differentiation to high-frequency noise is clearly demonstrated in this example. The high-frequency components present in the Fourier domain data are heavily weighted by the filter function compared with low-frequency data, and the signal-to-

---

[3] R. Bracewell, 'The Fourier Transform and its Applications', McGraw-Hill, New York, USA, 1965.

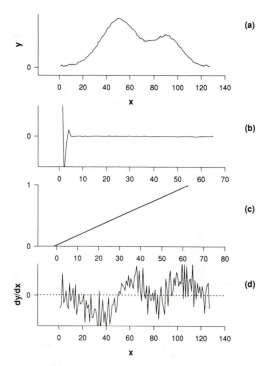

Figure 3 *Differentiation of spectra via the Fourier transform: (a) the original spectrum; (b) its Fourier transform; (c) the differential filter applied to the transform; (d) the resulting first derivative spectrum from the inverse transform*

noise ratio of the differential spectrum is severely degraded. This problem can be partly alleviated by combining a smoothing function along with the differential filter. In Figure 4(a), the differential transform is truncated and applied to low frequencies only. High frequencies are eliminated by the zero weighting of the function. The result of multiplying our transformed data by this new filter function is shown in Figure 4(b) and the resultant first-derivative spectrum in Figure 4(c). The effect of the extra smoothing function is evident if Figure 4(c) and Figure 3(d) are compared.

For many applications the digitization of a full spectrum provides far more data than is warranted by the spectrum's information content. An infrared spectrum, for example, is characterized as much by regions of no absorption as regions containing absorption bands, and most IR spectra can be reduced to a list of some 20–50 peaks. This represents such a dramatic decrease in dimensionality of the spectral data that it is not surprising that peak tables are commonly employed to describe spectra. The determination of spectral peak positions from digital data is relatively straightforward and the facility is offered on many commercial spectrometers. Probably the most common techniques for finding peak positions involve analysis of derivative data.

In Figure 5 a single Lorentzian function is illustrated along with its first,

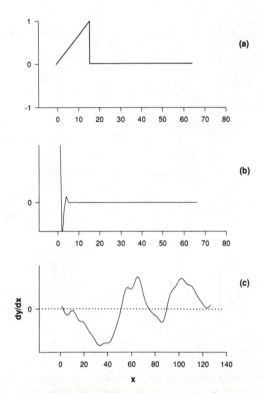

Figure 4 *Combining smoothing and differentiating in the frequency domain:* (a) *the truncation filter to remove high frequency, noise signals and provide the first derivative;* (b) *the transform of the spectrum form Figure 3(a) after application of the filter;* (c) *the resulting first derivative spectrum from the inverse transform*

second, third, and fourth derivatives with respect to energy. At peak positions the following conditions exist,

$$y' = 0 \qquad y'' < 0$$
$$y''' = 0 \qquad y'''' > 0 \qquad (6)$$

where y' is the first derivative, y'' the second, and so on.

Thus, the presence and location of a peak in a spectrum can be ascertained from a suitable subset of the rules expressed mathematically in Equation (6):[4]

Rule 1, a peak centre has been located if the first derivative value is zero and the second derivative value is negative, *i.e.*

$$(y' = 0) \quad \text{AND} \quad (y'' < 0), \text{ or}$$

[4] J.R. Morrey, *Anal. Chem.*, 1968, **40**, 905.

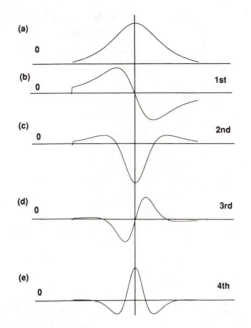

Figure 5 *A Lorentzian band (a), and its first (b), second (c), third (d), and fourth (e) derivatives*

Rule 2, a peak centre has been located if the third derivative is zero and the fourth derivative is positive, *i.e.*

$$(y''' = 0) \quad \text{AND} \quad (y'''' > 0)$$

Whereas Rule 2 is influenced less by adjacent, overlapping bands than Rule 1, it is affected more by noise in the data. In practice some form of Rule 1 is generally used. A peak-finding algorithm may take the following form:

Step 1: Convolute the spectrum with a suitable quadratic differentiating function until the computed central value changes sign.

Step 2: At this point of inflection compute a cubic, least-squares function. By numerical differentiation of the resultant equation determine the true position of zero slope (the peak position).

With any such algorithm it is necessary to specify some tolerance value below which any peaks are assumed to arise from noise in the data. The choice of window width for the quadratic differentiating function and the number of points about the observed inflection to fit the cubic model are selected by the user. These factors depend on the resolution of the recorded spectrum and the shape of the bands present. Results using a 15-point quadratic differentiating convolution function and a nine-point cubic fitting equation are illustrated in Figure 6.

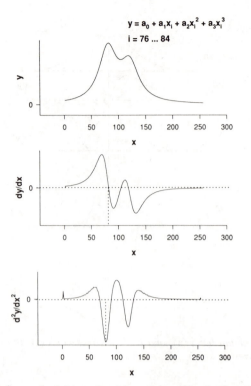

$$y = a_0 + a_1x_i + a_2x_i^2 + a_3x_i^3$$

$$i = 76 \ldots 84$$

Figure 6 *Results of a peak picking algorithm. At x = 80, the first derivative spectrum crosses zero and the second derivative is negative. A 9-point cubic least-squares fit is applied about this point to derive the coefficients of the cubic model. The peak position (dy/dx = 0) is calculated as occurring at x = 80.3*

3 Integration

Mathematically, integration is complementary to differentiation and comput-ing the integral of a function is a fundamental operation in data processing. It occurs frequently in analytical science in terms of determining the area under a curve, *e.g.* the integrated absorbance of a transient signal from a graphite furnace atomic absorption spectrometer. Many classic algorithms exist for approximating the area under a curve. We will briefly examine the more common with reference to the absorption profile illustrated in Figure 7. This envelope was generated from the model $y = (0.1x^3 - 1.1x^2 + 3x + 0.2)$. Its integral, between the limits $x = 0$ and $x = 6$, can be computed directly. The area under the curve is 8.400.

One of the simplest integration techniques to implement on a computer is the method of summing rectangles that each fit a portion of the curve, Figure 8(a). For $N + 1$ points in the interval $x_1, x_2 \ldots x_{N+1}$, we have N rectangles of width $(x_{i+1} - x_i)$ and height, h_{mi}, given by the value of the curve at the mid-point between x_i and x_{i+1}. The approximate area under the curve, A, between x_1 and x_{N+1} is therefore given by

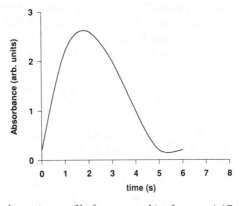

Figure 7 *The model absorption profile from a graphite furnace AAS study*

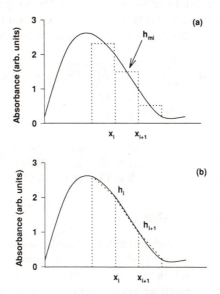

Figure 8 *The area under the AAS profile using* (a) *rectangular and* (b) *trapezoidal integration*

$$A = \sum_{i=1}^{N} (x_{i+1} - x_i) h_{mi} \tag{7}$$

As N gets larger, the width of each rectangle becomes smaller and the answer is more accurate:

for $N = 5,$ $A = 8.544$
$N = 10,$ $A = 8.436$
$N = 15,$ $A = 8.388$

A second method of approximating the integral is to divide the area under the curve into trapezoids, Figure 8(b). The area of each trapezoid is given by one-half the product of the width $(x_{i+1} - x_i)$ and the sum of the two sides, h_i and h_{i+1}. The area under the curve can be calculated from

$$A = \sum_{i=1}^{N} (x_{i+1} - x_i)(h_i + h_{i+1})/2 \tag{8}$$

For our absorption peak, the trapezoid method using different widths produces the following estimates for the integral:

$$
\begin{array}{lll}
\text{for} & N = 5, & A = 8.112 \\
& N = 10, & A = 8.328 \\
& N = 15, & A = 8.368
\end{array}
$$

In general the trapezoid method is inferior to the rectangular method.

A more accurate method can be achieved by combining the rectangular and trapezoid methods into the technique referred to as *Simpson's method*,[5]

$$A = \sum_{i=1}^{N} (x_{i+1} - x_i)(4h_{mi} + h_i + h_{i+1})/6 \tag{9}$$

For our absorption profile this gives

$$
\begin{array}{lll}
\text{for} & N = 5, & A = 8.400 \\
& N = 10, & A = 8.400 \\
& N = 15, & A = 8.400
\end{array}
$$

4   Combining Variables

Many analytical measures cannot be represented as a time-series in the form of a spectrum, but are comprised of discrete measurements, *e.g.* compositional or trace analysis. Data reduction can still play an important role in such cases. The interpretation of many multivariate problems can be simplified by considering not only the original variables but also *linear combinations* of them. That is, a new set of variables can be constructed each of which contains a sum of the original variables each suitably weighted. These linear combinations can be derived on an *ad hoc* basis or more formally using established mathematical techniques. Whatever the method used, however, the aim is to reduce the number of variables considered in subsequent analysis and obtain an improved representation of the original data. The number of variables measured is not reduced.

An important and commonly used procedure which generally satisfies these

[5] A.F. Carley and P.H. Morgan, 'Computational Methods in the Chemical Sciences', Ellis Horwood, Chichester, UK, 1989.

criteria is principal components analysis. Before this specific topic is examined it is worthwhile discussing some of the more general features associated with linear combinations of variables.

Linear Combinations of Variables

In order to consider the effects and results of combining different measured variables the data set shown in Table 3 will be analysed. Table 3 lists the mean values, from 11 determinations, of the concentration of each of 13 trace metals from 17 different samples of heart tissue.[6] The data in Table 3 indicate that the trace metal composition of cardiac tissue derived from different anatomical sites varies widely. However, it is not immediately apparent by visual examination of these raw data alone, what order, groups, or underlying patterns exist within the data.

The correlation matrix for the 13 variables is shown in Table 4 and, as is usual in multivariate data, some pairs of variables are highly correlated. Consider, in the first instance, the concentrations of chromium and nickel. We shall label these variables $X1$ and $X2$. These elements exhibit a mutual correlation of 0.90. A scatter plot of these data is illustrated in Figure 9. Also shown are projections of the points on to the $X1$ and $X2$ concentration axes, providing one-dimensional frequency distributions (as bar graphs) of the variables $X1$ and $X2$. It is evident from Figure 9 that a projection of the data points onto some other axis could provide this axis with a greater spread in terms of the frequency distribution. This single new variable or axis would contain more variance or potential information than either of the two original variables on their own. For example, a new variable, $X3$ could be identified which can be defined as the sum of the original variables, *i.e.*

$$X3 = a.X1 + b.X2 \qquad (10)$$

and its value for the 17 samples calculated. The values of a and b could be chosen arbitrarily such that, for example, $a = b$. Then, this variable would describe a new axis at an angle of 45° with the axes of Figure 9. The sample points can be projected on to this as illustrated in Figure 10.

As for the actual values of the coefficients a and b, the simplest case is described by $a = b = 1$, but any value will provide the same angle of projection and the same form of the distribution of data on this new line. In practice, it is usual to specify a particular linear combination referred as the *normalized linear combination* and defined by

$$a^2 + b^2 = 1 \qquad (11)$$

Normalization of the coefficients defining our new variable scales it to the range of values used to define the $X1$ and $X2$ axes of the original graph. In our

---

[6] W. Niedermeier, in 'Applied Atomic Spectroscopy', ed. E.L. Grove, Plenum Press, New York, USA, 1978, p. 219.

Table 3 *Heart-tissue trace metal data*

	Cu	Mn	Mo	Zn	Cr	Ni	Cs	Ba	Sr	Cd	Al	Sn	Pb
1 AO	38.4	5.1	1.9	300	2.3	1.8	3.9	9.9	5.4	7.1	31.0	15.2	8.8
2 MPA	47.6	3.8	2.1	774	1.2	1.7	2.3	4.0	3.7	9.7	13.7	15.9	8.8
3 RSCV	112.0	9.8	4.6	521	3.0	3.9	4.9	7.3	4.4	11.8	35.9	32.8	20.1
4 TV	52.7	3.7	3.2	358	3.0	2.7	3.7	6.6	3.6	17.6	28.6	18.5	25.2
5 MV	42.0	3.6	3.2	327	2.1	3.5	3.1	5.1	3.7	9.7	22.0	15.4	15.1
6 PV	102.0	21.9	5.9	744	7.6	8.4	47.3	54.4	16.4	33.9	280.0	36.9	55.0
7 AV	61.2	12.8	4.2	224	5.7	5.2	21.9	31.7	7.8	13.8	126.0	25.2	32.5
8 RA	140.0	6.4	3.3	429	2.2	2.8	2.0	2.3	2.3	7.5	5.0	33.9	12.2
9 LAA	137.0	7.9	3.2	353	0.7	2.8	1.1	0.7	1.2	7.9	6.5	33.5	9.5
10 RV	163.0	6.7	3.8	433	2.0	2.2	0.8	0.7	1.6	7.5	6.3	38.9	6.7
11 LV	171.0	7.2	4.6	396	1.7	2.8	0.7	0.8	1.5	7.3	6.4	44.2	7.6
12 LV-PM	170.0	7.8	5.3	330	1.8	2.6	0.9	0.8	2.1	6.8	7.6	41.6	9.3
13 IVS	171.0	6.8	4.6	248	1.1	2.8	1.4	0.9	1.5	7.0	9.5	39.6	10.3
14 CR	160.0	6.9	4.1	493	1.5	2.9	1.3	1.6	1.9	7.7	17.8	35.6	12.5
15 SN	145.0	6.6	4.2	548	1.7	2.9	1.1	1.5	2.2	7.7	8.6	31.5	12.4
16 AVN + B	144.0	6.5	4.2	284	1.3	2.5	1.3	1.3	3.3	6.5	5.4	32.8	7.5
17 LBB	164.0	7.7	5.4	449	1.5	2.8	1.5	1.2	2.1	7.6	12.8	39.7	12.6

AO Aorta; MPA Main pulmonary artery; RSCV Right superior vena cava; TV Tricuspid valve; MV Mitral valve; PV Pulmonary valve; AV Aortic valve; RA Right atrium; LAA Left atrial appendage; RV Right ventricle; LV Left ventricle; LV-PM Left ventricle, muscle; IVS Interventricular septum; CR Crista supraventricularis; SN sinus node; AVN + B Atrioventricular node; LBB Left bundle branch.

Table 4 *The matrix of correlations between the analytes determined from heart-tissue data*

	Cu	Mn	Mo	Zn	Cr	Ni	Cs	Ba	Sr	Cd	Al	Sn	Pb
Cu	1												
Mn	0.08	1											
Mo	0.61	0.65	1										
Zn	-0.14	0.33	0.08	1									
Cr	-0.36	0.83	0.36	0.25	1								
Ni	-0.13	0.93	0.57	0.34	0.90	1							
Cs	-0.27	0.91	0.40	0.38	0.94	0.95	1						
Ba	-0.38	0.87	0.32	0.34	0.96	0.92	0.99	1					
Sr	-0.42	0.83	0.28	0.42	0.93	0.88	0.97	0.98	1				
Cd	-0.34	0.78	0.34	0.48	0.88	0.88	0.92	0.89	0.91	1			
Al	-0.28	0.91	0.40	0.39	0.94	0.94	0.99	0.99	0.97	0.93	1		
Sn	0.94	0.36	0.77	-0.07	-0.07	0.14	0.01	-0.10	-0.16	-0.08	-0.01	1	
Pb	0.34	0.83	0.41	0.37	0.95	0.94	0.95	0.94	0.92	0.96	0.95	-0.07	1

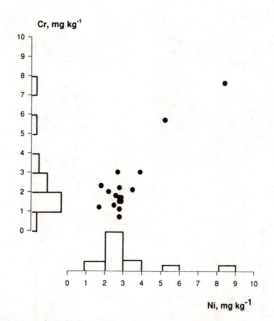

Figure 9   *Chromium and nickel concentration scatter plot from heart tissue data. The distribution of concentration values for each element is shown as a bar graph on their respective axes*

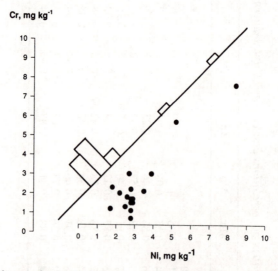

Figure 10   *A 45° line on the Cr–Ni data plot with the individual sample points projected on to this line*

example, this implies $a = b = 1/\sqrt{2}$. The variance of $X3$ derived from substituting a and b into Equation 10 for the concentration of chromium and nickel for each of the 17 samples is 5.22 compared with $\sigma^2 = 3.07$ and $\sigma^2 = 2.43$ for $X1$ and $X2$ respectively. Thus $X3$ does indeed contain more potential information than either $X1$ or $X2$.

This reorganization or partitioning of variance associated with individual variates can be formally addressed as follows.

For any linear combination of variables defining a new variable X given by

$$X = a_1 x_2 + a_2 x_2 + \ldots + a_n x_n \tag{12}$$

The variance, s_x^2, of the new variable can be calculated from

$$s_x^2 = \sum_{j=1}^{n} \sum_{k=1}^{n} a_j . a_k \, Cov_{jk} \tag{13}$$

which, from the definition of covariance, can be rewritten as

$$s_x^2 = \sum_{j=1}^{n} a_j^2 . s_j^2 + \sum_{j=1}^{n} \sum_{k=j+1}^{n} a_j . s_j . a_k . s_k . r_{jk} \tag{14}$$

where r_{jk} is the correlation coefficient between variables x_j and x_k.

It should be noted that for statistically independent variables, $r_{jk} = 0$ and Equation (14) reduces to the more common equation stating that the variance of a sum of variables is equal to the sum of the variances for each variable.

The calculated value for the variance of our new variable $X3$ confirms that there is an increased spread of the data on the new axis. As well as this algegbraic notation, it is worth pointing out that the coefficients of the normalized linear combination may be represented by the trigonometric identities

$$a = \cos \alpha$$
$$b = \sin \alpha \tag{15}$$

where α is the angle between the projection of the new axis and the original ordinate axis. If $\alpha = 45°$, then $a = b = 1/\sqrt{2}$, the normalized coefficients as derived from Equation (11). This trigonometric relationship is often employed in determining different linear combinations of variables and is used in many principal component algorithms.

Values of α, or a and b, employed in practice depend on the aims of the data analysis. Different linear combinations of the same variables will produce new variables with different attributes which may be of interest in studying different problems.[7] The linear combination which produces the greatest separation between two groups of data samples is appropriate in supervised pattern

---

[7] B. Flury and H. Riedwye, 'Multivariate Statistics: A Practical Approach', Chapman and Hall, London, UK, 1984.

recognition. This forms the basis of linear discriminant analysis, a topic that will be discussed in Chapter 5. Considering our samples or objects as a single group or cluster, we may wish to determine the minimum number of normalized linear combinations having the greatest proportion of the total variance, in order to reduce the dimensionality of the problem. This is the task of principal components analysis and is treated in the next section.

Principal Components Analysis

The aims of performing a principal components analysis (PCA) on multivariate data are basically two-fold. Firstly, PCA involves rotating and transforming the original, n, axes each representing an original variable into new axes. This transformation is performed in a way so that the new axes lie along the directions of maximum variance of the data with the constraint that the axes are *orthogonal*, *i.e.* the new variables are uncorrelated. It is usually the case that the number of new variables, p, needed to describe most of the sample data variance is less than n. Thus PCA affords a method and a technique to reduce the dimensionality of the parameter space. Secondly, PCA can reveal those variables, or combinations of variables, that determine some inherent structure in the data and these may be interpreted in chemical or physico-chemical terms.

As in the previous section, we are interested in linear combinations of variables, with the goal of determining that combination which best summarizes the n-dimensional distribution of data. We are seeking the linear combination with the largest variance, with normalized coefficients applied to the variables used in the linear combinations. This axis is the so-called first *principal axis* or first *principal component*. Once this is determined, then the search proceeds to find a second normalized linear combination that has most of the remaining variance and is uncorrelated with the first principal component. The procedure is continued, usually until all the principial components have been calculated. In this case, $p = n$ and a selected subset of the principal components is then used for further analysis and for interpretation.

Before proceeding to examine how principal components are calculated, it is worthwhile considering further a graphical interpretation of their structure and characteristics.[8] From our heart tissue, trace metal data, the variance of chromium concentration is 3.07, the variance of nickel concentration is 2.43, and their covariance is 2.47. This variance–covariance structure is represented by the variance–covariance matrix,

$$Cov_{X1,X2} = \begin{array}{c} \\ X1 \\ X2 \end{array} \begin{array}{cc} X1 & X2 \\ 3.07 & 2.47 \\ 2.47 & 2.43 \end{array} \tag{16}$$

As well as this matrix form, this structure can also be represented diagrammatically as shown in Figure 11. The variance of the chromium data is represented by a line along the $X1$ axis with a length equal to the variance of $X1$.

[8] J.C. Davis, 'Statistics in Data Analysis in Geology', J. Wiley and Sons, New York, USA, 1973.

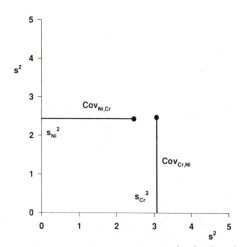

Figure 11 *A bivariate variance–covariance matrix may be displayed graphically*

Since the concentration of chromium is correlated with the concentration of nickel, $X1$ values vary with variable $X2$ axis. The length of this line is equal to the covariance of $X1$ with $X2$ and represents the degree of interaction or *colinearity* between the variables. In a similar manner, the variance and covariance of $X2$ are drawn along and from the second axis. For a square (2×2) matrix, these elements of the variance–covariance matrix lie on the boundary of an ellipse, the centre of which is the origin of the co-ordinate system. The slope of the major axis is the *eigenvector* associated with the first principal component, and its corresponding *eigenvalue* is the length of this major axis, Figure 12. The second principal component is defined by the second eigenvector and eigenvalue. It is represented by the minor axis of the ellipse and is orthogonal, $90°$, to the first principal component. For a 3×3 variance–covariance matrix the elements lie on the surface of a three-dimensional ellipsoid. For larger matrices still, higher-dimensional elliptical shapes apply and can only be imagined. Fortunately the mathematical operations deriving and defining these components remain the same whatever the dimensionality.

Thus, principal components can be defined as the eigenvectors of a variance–covariance matrix. They provide the direction of new axes (new variables) on to which data can be projected. The size, or length, of these new axes containing our projected data is proportional to the variance of the new variable.

How do we calculate these eigenvectors and eigenvalues? In practice the calculations are always performed on a computer and there are many algorithms published in mathematical and chemometric texts. For our purposes, in order to illustrate their derivation, we will limit ourselves to bivariate data and calculate the eigenvectors manually. The procedure adopted largely follows that of Davis[8] and Healy.[9]

[9] M.J.R. Healy, 'Matrices for Statistics', Oxford University Press, Oxford, UK, 1986.

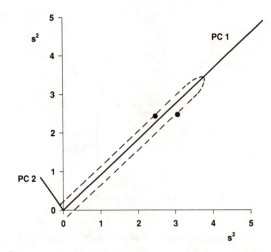

Figure 12 *The elements of a bivariate variance–covariance matrix lie on the boundary defined by an ellipse. The major axis of the ellipse represents the first principal component, and its minor axis the second principal component*

Consider a set of simultaneous equations, expressed in matrix notation,

$$A.x = \ell.x \qquad (17)$$

which simply states that matrix A, multiplied by vector x, is equal to some constant, the eigenvalue ℓ, multiplied by x. To determine these eigenvalues, Equation (17) can be rewritten as

$$A.x - \ell.x = 0 \qquad (18)$$

or

$$(A - \ell.I).x = 0 \qquad (19)$$

where I is the identity matrix, which for a 2×2 matrix is

$$I = \begin{bmatrix} 1 & 0 \\ 0 & 1 \end{bmatrix} \qquad (20)$$

If x is not 0 then the determinant of the coefficient matrix must be zero, *i.e.*

$$|A - \ell.I| = 0 \qquad (21)$$

For our experimental data with $X1$ and $X2$ representing chromium and nickel concentrations, and $A = Cov$, then

$$|\boldsymbol{Cov} - \ell.\boldsymbol{I}| = \begin{vmatrix} s_{x1}^2 - \ell & Cov \\ Cov & s_{x2}^2 - \ell \end{vmatrix} = 0 \qquad (22)$$

where *Cov* is the covariance between $X1$ and $X2$. Expanding Equation (22) gives the quadratic equation,

$$(s_{x1}^2 - \ell)(s_{x2}^2 - \ell) - Cov^2 = 0 \qquad (23)$$

and substituting the values for our Cr and Ni data,

$$(3.07 - \ell)(2.43 - \ell) - 2.47^2 = 0 \qquad (24)$$

which simplifies to

$$\ell^2 - 5.5\ell + 1.36 = 0 \qquad (25)$$

This is a simple quadratic equation providing two *characteristic roots* or eigenvalues, *viz.* $\ell_1 = 5.24$ and $\ell_2 = 0.26$.

As a check in our calculations, the sum of the eigenvalues should be equal to the sum of diagonal elements, the *trace*, of the original matrix (*i.e.* $3.07 + 2.43 = 5.24 + 0.26$).

Associated with each eigenvalue (the length of the new axis in our geometric model) is a *characteristic vector*, the eigenvector, $\boldsymbol{v} = [v_1, v_2]$ defining the slope of the axis. Our eigenvalues, ℓ, were defined as arising from a set of simultaneous equations, Equation (19), which can now be expressed, for a 2×2 matrix, as,

$$\begin{bmatrix} A_{11} - \ell_1 & A_{12} \\ A_{21} & A_{22} - \ell_1 \end{bmatrix} \cdot \begin{bmatrix} X_1 \\ X_2 \end{bmatrix} = \begin{bmatrix} 0 \\ 0 \end{bmatrix} \qquad (26)$$

and the elements of x are the eigenvectors associated with the first eigenvalue, ℓ_1. For our 2×2, Ni–Cr variance–covariance data, substitution into (26) leads to

$$\begin{bmatrix} s_{x1}^2 - \ell_1 & Cov \\ Cov & s_{x2}^2 - \ell_1 \end{bmatrix} \cdot \begin{bmatrix} v1_1 \\ v1_2 \end{bmatrix} = 0 \qquad (27)$$

$$\begin{bmatrix} s_{x1}^2 - \ell_2 & Cov \\ Cov & s_{x2}^2 - \ell_2 \end{bmatrix} \cdot \begin{bmatrix} v2_1 \\ v2_2 \end{bmatrix} = 0 \qquad (28)$$

with $v1_1$ and $v1_2$ as the eigenvectors associated with the first eigenvalue, and $v2_1$ and $v2_2$ defining the slope of the second eigenvalue.

Solving these equations gives

$$v1 = [0.751, 0.660] \qquad (29)$$

which defines the slope of the major axis of the ellipse (Figure 12), and

$$v2 = [-0.660, 0.751] \tag{30}$$

which is perpendicular to v1 and is the slope of the ellipse's minor axis.

Having determined the orthogonal axes or principal components of our bivariate data, it remains to undertake the projections of the data points on to the new axes. For the first principal components $PC1$,

$$PC1_i = 0.751X1_i + 0.660X2_i \tag{31}$$

and for the second principal component $PC2$,

$$PC2_i = -0.660X1_i + 0.751X2_i \tag{32}$$

Thus, the elements of the eigenvectors become the required coefficients for the original variables, and are referred to as *loadings*. The individual elements of the new variables ($PC1$ and $PC2$) are derived from $X1$ and $X2$ and are termed the *scores*.[10,11] The principal components scores for the chromium and nickel data are given in Table 5.

The total variance of the original nickel and chromium data is $3.07 + 2.43 = 5.5$ with $X1$ contributing 56% of the variance and $X2$ contribut-

Table 5 *The PC scores for chromium and nickel concentrations*

Sample	PC1	PC2
AO	2.91	−0.17
MPA	2.02	0.48
RSCV	4.82	0.94
TV	4.03	0.04
MV	3.89	1.24
PV	11.25	1.28
AV	7.71	0.14
RA	3.50	0.65
LAA	2.37	1.64
RV	2.95	0.33
LV	3.12	0.98
LV-PM	3.06	0.76
IVS	2.67	1.37
CR	3.04	1.18
SN	3.19	1.05
AVN + B	2.62	1.02
LBB	2.97	1.11

[10] B.F.J. Manly, 'Multivariate Statistical Methods: A Primer', Chapman and Hall, London, UK, 1986.
[11] R.E. Aries, D.P. Lidiard, and R.A. Spragg, *Chem. Br.*, 1991, **27**, 821.

ing the remaining 44%. The calculated eigenvalues are the lengths of the two principal axes and represent the variance associated with each new variable, *PC*1 and *PC*2. The first principal component, therefore, contains 5.24/5.50 or more than 95% of the total variance, and the second principal component less than 5%, 0.26/5.50. If it were necessary to reduce the display of our original bivariate data to a single dimension using only one variable, say chromium concentration, then a loss of 44% of the total variance would ensue. Using the first principal component, however, and optimally combining the two variables, only 5% of the total variance would be missing.

We are now in a position to return to the complete set of trace element data in Table 3 and apply principal components analysis to the full data matrix. The techniques described and used in the above example to extract and determine the eigenvalues and eigenvectors for two variables can be extended to the more general, multivariate case but the procedure becomes increasingly difficult and arithmetically tedious with large matrices. Instead, the eigenvalues are usually found by matrix manipulation and iterative approximation methods using appropriate computer software. Before such an analysis is undertaken, the question of whether to transform the original data should be considered. Examination of Table 3 indicates that the variates considered have widely differing means and standard deviations. Rather than standardizing the data, since they are all recorded in the same units, one other useful transformation is to take logarithms of the values. The result of this transformation is to scale all the data to a more similar range and reduce the relative effects of the more concentrated metals. Having performed the log-transformation on our data, the results of performing PCA on all 13 for the 17 samples are as given in Table 6.

According to the eigenvalue results present in Table 6(b), and displayed in the *scree plot* of Figure 13, over 84% of the total variance in the original data

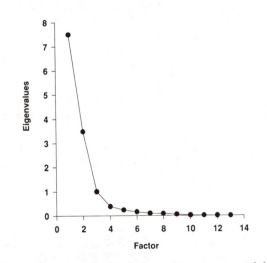

Figure 13 *An eigenvalue, scree plot for the heart-tissue trace metal data*

Table 6 Results of principal components analysis on the logarithms of the trace metals concentration data

(a) Eigenvalues

	PC1	PC2	PC3	PV4	PC5	PC6	PC7	PC8	PC9	PC10	PC11	PC12	PC13
Variance	7.49	3.47	0.99	0.38	0.23	0.15	0.09	0.08	0.05	0.02	0.01	0.000	0.001
% Variance	57.7	26.7	7.7	2.9	1.8	1.2	0.7	0.6	0.4	0.1	0.1	0.0	0.0
Cumulative % contribution	57.7	84.4	92.1	95.0	96.8	98.0	98.7	99.4	99.8	99.9	100	100	100

(b) Eigenvectors

	PC1	PC2	PC3	PV4	PC5	PC6	PC7	PC8	PC9	PC10	PC11	PC12	PC13
Log Cu	−0.17	0.47	0.08	0.08	0.07	0.16	−0.07	−0.17	0.35	0.47	−0.43	−0.22	0.35
Log Mn	0.20	0.39	−0.05	0.57	−0.30	0.12	0.14	0.27	−0.16	−0.54	−0.17	0.001	0.13
Log Mo	0.63	0.49	−0.04	−0.33	0.26	−0.52	0.36	−0.009	0.03	−0.30	0.20	0.02	0.26
Log Zn	0.10	0.01	0.95	0.10	0.12	−0.11	−0.11	0.10	−0.08	−0.03	0.08	−0.05	−0.006
Log Cr	0.33	0.04	−0.19	−0.03	0.74	0.26	−0.30	0.20	−0.23	−0.08	−0.006	−0.21	0.01
Log Ni	0.29	0.27	−0.05	−0.24	−0.30	−0.26	−0.50	−0.26	−0.45	0.28	−0.06	−0.002	−0.08
Log Cs	0.36	−0.04	−0.07	0.15	−0.20	0.002	−0.05	−0.12	0.35	0.11	0.58	−0.55	0.10
Log Ba	0.35	−0.14	−0.03	0.17	0.04	−0.05	−0.16	0.08	0.19	0.19	0.14	0.68	0.48
Log Sr	0.34	−0.11	−0.01	0.33	0.27	−0.31	0.24	−0.50	0.19	0.10	−0.32	0.04	−0.36
Log Cd	0.33	0.19	0.17	−0.36	−0.07	0.59	0.41	−0.38	−0.17	0.005	−0.02	0.07	0.16
Log Al	0.35	−0.03	−0.05	0.06	−0.18	−0.11	0.44	0.63	−0.21	0.38	−0.14	−0.09	−0.11
Log Sn	−0.09	0.51	0.01	0.14	0.08	0.26	0.38	0.05	0.08	0.22	0.42	0.34	−0.53
Log Pb	0.34	0.07	0.04	−0.41	−0.15	0.06	−0.22	0.26	0.55	−0.31	−0.26	0.07	−0.29

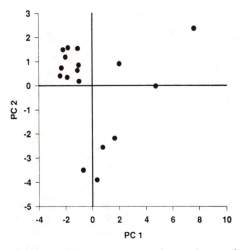

Figure 14 *Scatter plot of the 17 heart-tissue samples on the standardized first two principal components from the trace metal data*

can be accounted for by the first two principal components. The transformation of the 13 original variables to two new linear combinations represents considerable reduction of the data presented whilst retaining much of the original information. A scatter plot of the first two principal components scores is shown in Figure 14 and patterns to the samples according to the distribution of the trace metals in the data are evident. Three tissues, the pulmonary valve, aortic valve, and the right superior vena cava, constitute unique groups of one tissue each, well distinguished from the rest. The aorta, main pulmonary artery, mitral, and tricuspid valves constitute a cluster of four tissue types. Finally, there is a group of ten tissues derived from the myocardium. A more detailed analysis and discussion of this data set is presented by Niedermeier.[6]

As well as being used with discrete analytical data, such as the trace metal concentrations discussed above, principal components analysis has been extensively employed on digitized spectral profiles.[12] A simple example will illustrate the basis of these applications. Infrared spectra of 21 samples of acrylic, PVC, styrene, and nylon polymers, as thin films, were recorded in the range 4000–600 cm^{-1}. Each spectrum was normalized on the most intense absorption band to remove film thickness effects, and reduced to 216 discrete values by signal averaging. The resulting 21 × 216 data matrix was subject to principal components analysis. The resulting eigenvalues are illustrated in the scree plot of Figure 15, and the first three principal components account for more than 91% of the total variance in the original spectra. A scatter plot of the polymer data loaded on to these three components is shown in Figure 16. It is evident from this plot that these three components are sufficient to provide effective

12 I.A. Cowe and J.W. McNicol, *Appl. Spectrosc.*, 1985, **39**, 257.

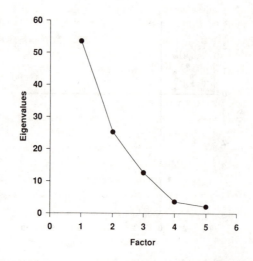

Figure 15 *A scree plot for the eigenvalues derived from the IR spectra of 21 polymers*

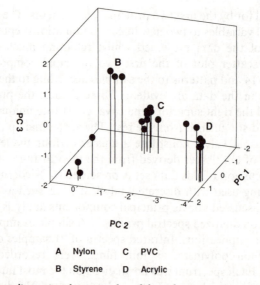

| A | Nylon | C | PVC |
| B | Styrene | D | Acrylic |

Figure 16 *A three-dimensional scatter plot of the polymer spectra projected on to the first three principal components*

clustering of the samples with clear separation between the groups and types of polymer. The first component, *PC*1, forms an axis which would allow the partitioning between acrylic polymer and other samples. *PC*2 provides for two partitions; both the nylon and PVC polymers are separated from the styrene and acrylic polymers. *PC*3 allows the separation between styrenes and others.

Examination of the principal component loadings, the eigenvectors, as functions of wavelength, *i.e.* spectra of loadings, highlights the weights given to each spectral point in each of the original spectra, Figure 17. It can be seen from these 'spectra' that where a partition between sample types if formed, the majority of absorption bands in the corresponding spectra receive strong positive or negative weighting. *PC*2 produces two partitions and, Figure 17(b), the bands in nylon spectra receive positive weightings and bands in PVC spectra, Figure 17(c), have negative weightings.

The power of principal components analysis is in providing a mathematical transformation of our analytical data to a form with reduced dimensionality. From the results, the similarity and difference between objects and samples can often be better assessed and this makes the technique of prime importance in chemometrics. Having introduced the methodology and basics here, future chapters will consider the use of the technique as a data preprocessing tool.

Factor Analysis

The extraction of the eigenvectors from a symmetric data matrix forms the basis and starting point of many multivariate chemometric procedures. The way in which the data are preprocessed and scaled, and how the resulting vectors are treated, has produced a wide range of related and similar techniques. By far the most common is principal components analysis. As we have seen, PCA provides n eigenvectors derived from a $n \times n$ dispersion matrix of variances and covariances, or correlations. If the data are standardized prior to eigenvector analysis, then the variance–covariance matrix becomes the correlation matrix [see Equation (25) in Chapter 1, with $s_1 = s_2$]. Another technique, strongly related to PCA, is factor analysis.[13]

Factor analysis is the name given to eigen analysis of a data matrix with the intended aim of reducing the data set of n variables to a specified number, p, of fewer linear combination variables, or *factors*, with which to describe the data. Thus, p is selected to be less than n and, hopefully, the new data matrix will be more amenable to interpretation. The final interpretation of the meaning and significance of these new factors lies with the user and the context of the problem.

A full description and derivation of the many factor analysis methods reported in the analytical literature is beyond the scope of this book. We will limit ourselves here to the general and underlying features associated with the technique. A more detailed account is provided by, for example, Hopke[14,15] and others.[16-19]

13 D. Child, 'The Essentials of Factor Analysis', 2nd Edn, Cassel Educational, London, UK, 1990.
14 P.K. Hopke, in 'Methods of Environmental Data Analysis', ed. C.N. Hewitt, Elsevier, Essex, UK, 1992.
15 P.K. Hopke, *Chemomet. Intell. Lab. Systems*, 1989, **6**, 7.
16 E. Malinowski, 'Factor Analysis in Chemistry', J. Wiley and Sons, New York, USA, 1991.
17 T.P.E. Auf der Heyde, *J. Chem. Ed.*, 1983, **7**, 149.
18 G.L. Ritter, S.R. Lowry, T.L. Isenhour, and C.L. Wilkins, *Anal. Chem.*, 1976, **48**, 591.
19 E. Malinowski and M. McCue, *Anal. Chem.*, 1977, **49**, 284.

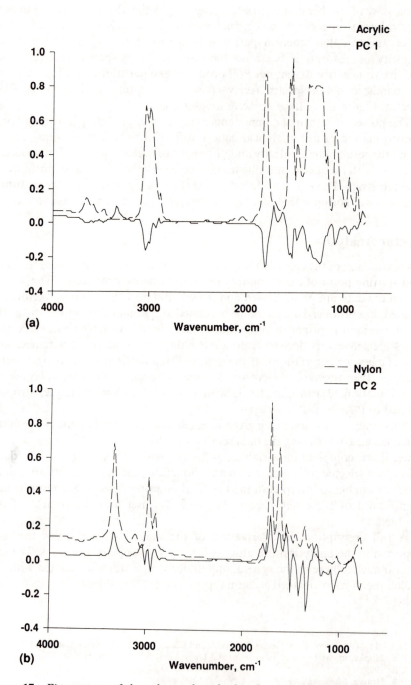

Figure 17 *Eigenvectors of the polymer data displayed as a function of wavelength and
compared with typical spectra: (a) the first principal component and a spectrum
of an acrylic sample; (b) the second PC and a nylon spectrum; (c) the second
PC and a PVC spectrum; and (d) the third PC and a polystyrene spectrum*

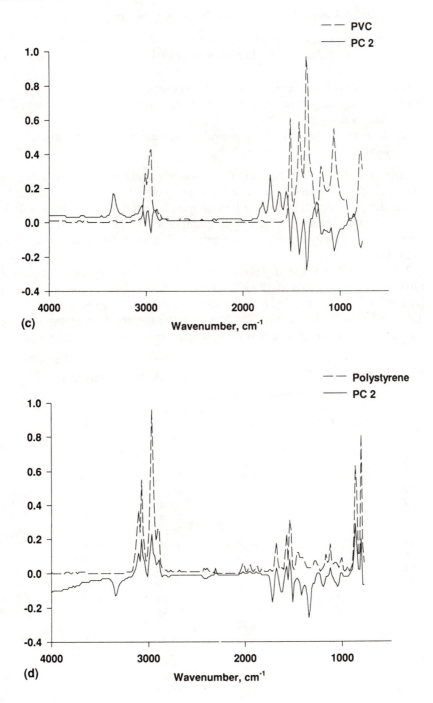

(c)

Wavenumber, cm⁻¹

(d)

Wavenumber, cm⁻¹

The principal steps in performing a factor analysis are,

(a) preprocessing of the raw data matrix,
(b) computing the symmetric matrix of covariances or correlations, *i.e.* the *dispersion matrix*,
(c) extracting the eigenvalues and eigenvectors,
(d) selecting the appropriate number of factors with which to describe the data, and
(e) rotating these factors to provide a meaningful interpretation of the factors.

Steps (a) to (c) are as for principal components analysis. However, as the final aim is usually to interpret the results of the analysis in terms of chemical or spectroscopic properties, the method adopted at each step should be selected with care and forethought. A simple example will serve to illustrate the principles of factor analysis and the application of some of the options available at each stage.

Table 7 provides the digitized mass spectra of five cyclohexane/hexane mixtures, each recorded at 17 m/z values and normalized to the most intense, parent ion.[20] These spectra are illustrated in Figure 18. Presented with these data, and in a 'real' situation not knowing the composition of the mixtures, our first task is to determine how many discrete components contribute to these

Table 7 *Normalized MS data for cyclohexane and hexane mixtures*

| | | | % Cyclohexane | | |
| | 90 | 80 | 50 | 20 | 10 |
m/z	A	B	C	D	E
27	13.79	19.05	20.80	28.30	24.55
29	12.93	15.87	26.40	44.04	38.18
39	17.24	17.46	20.00	20.02	18.18
40	4.31	4.76	4.00	3.00	3.64
41	55.17	63.49	14.40	91.09	80.00
42	29.31	29.37	36.80	47.05	41.82
43	21.55	26.19	49.60	74.07	71.82
44	1.72	1.59	1.60	3.00	1.82
54	5.17	4.76	4.00	3.00	1.82
55	31.90	34.13	28.00	24.02	10.00
56	100.00	100.00	100.00	100.00	73.36
57	19.83	25.40	58.40	96.10	100.00
69	29.31	26.98	21.60	16.02	9.09
83	5.17	4.76	4.00	4.00	3.64
84	70.69	68.25	58.40	36.04	18.18
85	6.90	6.35	4.80	5.00	2.73
86	3.45	4.76	12.80	24.02	22.73
Mean	25.20	26.66	30.92	36.33	30.86

[20] R.W. Rozett and E.M. Petersen, *Anal. Chem.*, 1975, **47**, 1301.

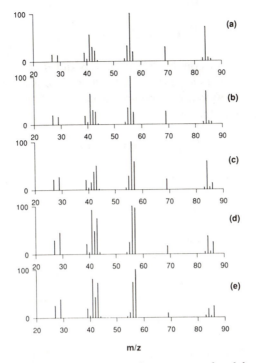

Figure 18   *The mass spectra recorded from five mixtures of cyclohexane and hexane*

spectra, *i.e.* how many components are in the mixtures. We can then attempt to identify the nature or source of each extracted component. These are the aims of factor analysis.

Before we can compute the eigenvectors associated with our data matrix, we need to select appropriate, if any, preprocessing methods for the data, and the form of the dispersion matrix. Specifically, we can choose to generate a covariance matrix or a correlation matrix from the data. Each of these could be derived from the original, origin-centred data or from transformed, mean-centred data. In addition, we should bear in mind the aim of the analysis and decide whether the variables for numerical analysis are the m/z values or the composition of the sample mixtures themselves. Thus we have eight options in forming the transformed, symmetric matrix for extracting eigenvectors. We can form a 5×5 covariance, or correlation, matrix on the origin- or mean-centred compositional values. Alternatively, a 17×17 covariance, or correlation, matrix can be formed from origin- or mean-centred m/z values.

Each of these transformations can be expressed in matrix form as a transform of the data matrix X to a new matrix Y followed by calculating the appropriate dispersion matrix, C (the variance–covariance, or correlation matrix). The relevant equations are

$$Y = XA + B \tag{33}$$

and

$$C = Y^T.Y/(n-1) \tag{34}$$

The nature of C depends on the definition of A and B. A is a scaling diagonal matrix; only the diagonal elements need be defined. B is a centring matrix in which all elements in any one column are identical.

For covariance about the origin,

$$a_{jj} = 1 \quad \text{and} \quad b_{ij} = 0 \tag{35}$$

For covariance about the mean,

$$a_{jj} = 1 \quad \text{and} \quad b_{ij} = -\bar{x}_j \tag{36}$$

For correlation about the origin,

$$a_{jj} = \left(\frac{1}{n}\sum_{i=1}^{n} x_{ij}\right)^{-1/2}$$

$$b_{ij} = 0 \tag{37}$$

For correlation about the mean,

$$a_{jj} = \left(\frac{1}{n-1}\sum_{i=1}^{n}(x_{ij} - \bar{x}_j)^2\right)^{-1/2}$$

$$b_{ij} = \bar{x}_j.a_{jj} \tag{38}$$

where $\bar{x}_j$ is the mean value of column j from the data matrix.

Mean-centring is a common pre-processing transformation as it provides data which are symmetric about a zero mean. It is recommended as a pre-processing step in many applications. This is not necessary here with our mass spectra, however, as the intensity scale has a meaningful zero.

As the analytical data are all in the same units and cover a similar range of magnitude, standardization is not required either and the variance–covariance matrix will be used as the dispersion matrix.

The final decision to be made is to whether to operate on the m/z values or the samples (actually the mixture compositions) as the analytical variables. It is a stated aim of our factor analysis to determine some physical meaning of the derived factors. We do not wish simply to perform a mathematical transformation to reduce the dimensionality of the data, as would be the case with principal components analysis.

We will proceed, therefore, with an eigenvector analysis of the 5×5 covariance matrix obtained from zero-centred object data. This is referred to as *Q-mode* factor analysis and is complementary to the scheme illustrated pre-

Table 8 *The variance–covariance matrix for the MS data and the eigenvalues and eigenvectors extracted from this*

			Covariance matrix		
	A	B	C	D	E
A	726.62				
B	726.04	734.27			
C	683.14	713.48	808.63		
D	594.23	655.34	890.97	1157.49	
E	421.65	485.71	754.65	1065.58	1025.79

			Eigenvalues		
Factor	1	2	3	4	5
Eigenvalue	3744.08	703.56	2.84	1.25	1.07
Cumulative % contribution	84.08	99.88	99.95	99.98	100.00

			Eigenvectors		
	$F(I)$	$F(II)$	$F(III)$	$F(IV)$	$F(V)$
A	0.368	0.557	−0.424	0.607	−0.075
B	0.389	0.486	0.401	−0.462	−0.488
C	0.461	0.121	−0.193	−0.434	0.740
D	0.533	−0.359	0.605	−0.446	0.146
E	0.464	−0.557	−0.505	−0.176	−0.434

viously with principal components analysis. In the earlier example the dispersion matrix was formed between the measured trace metal variables, and the technique is sometimes referred to as *R-mode* analysis. For the current MS data, processing by R-mode analysis would involve the data being scaled along each m/z column and information about relative peak sizes in any single spectrum would be destroyed. In Q-mode analysis, any scaling is performed within a spectrum and the mass fragmentation pattern for each sample is preserved.

The variance–covariance matrix of the mass spectra data is presented in Table 8, along with the results of computing the eigenvectors and eigenvalues from this matrix. In factor analysis we assume that any relationships between our samples from within the original data set can be represented by p mutually uncorrelated underlying factors. The value of p is usually selected to be much less than the number of original variables. These p underlying factors, or new variables, are referred to as *common factors* and may be amenable to physical interpretation. The remaining variance not accounted for by the p factors will be contained in a *unique factor* and may be attributable, for example, to noise in the system.

The first requirement is to determine the appropriate value of p, the number of factors necessary to describe the original data adequately. If p cannot be specified then the partition of total variance between common and unique factors cannot be determined. For our simple example with the mass spectra data it appears obvious that $p = 2$, *i.e.* there are two common factors which we

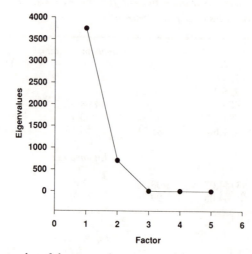

Figure 19   *The scree plot of the eigenvalues extracted from the MS data*

may interpret as being due to two components in the mixtures. The eigenvalues drop markedly from the second to the third value, as can be seen from Table 8 and Figure 19. The first two factors account for more than 99% of the total variance. The choice is not always so clear, however, and in the chemometrics literature a number of more objective functions have been described to select appropriate values of p.[14]

The eigenvectors in Table 8 have been normalized, *i.e.* each vector has unit length, *viz.*,

$$(0.368)^2 + (0.389)^2 + (0.461)^2 + (0.533)^2 + (0.464)^2 = 1 \qquad (39)$$

To perform factor analysis, the eigenvectors should be converted so that each vector length represents the magnitude of the eigenvalue. This conversion is achieved by multiplying each element in the normalized eigenvector matrix by the square root of the corresponding eigenvalue. From Table 8, the variance associated with the first factor is its eigenvalue, 3744.08, and the first eigenvector is converted to the first factor by multiplying by $\sqrt{(3744.08)}$, *viz.*

$$\text{Factor } 1 = \begin{pmatrix} 0.368 & \sqrt{(3744.08)} \\ 0.389 & \sqrt{(3744.08)} \\ 0.461 & \sqrt{(3744.08)} \\ 0.533 & \sqrt{(3744.08)} \\ 0.464 & \sqrt{(3744.08)} \end{pmatrix} = \begin{pmatrix} 22.52 \\ 23.82 \\ 28.24 \\ 32.64 \\ 29.40 \end{pmatrix} \qquad (40)$$

The elements in each of the factors are the factor loadings, and the complete factor loading matrix for our MS data is given in Table 9. This conversion has not changed the orientation of the factor axes from the original eigenvectors

Table 9 *The factor loading matrix from a Q-mode analysis of the MS data*

	F(I)	F(II)	F(III)	F(IV)	F(V)
A	22.52	14.78	− 0.71	0.68	− 0.08
B	23.82	12.88	0.68	− 0.52	− 0.51
C	28.24	3.20	− 0.32	− 0.48	0.77
D	32.64	− 9.51	1.02	0.50	0.15
E	28.40	− 14.78	− 0.85	− 0.20	− 0.45

but has simply changed their absolute magnitude. The lengths of each vector are now equal to the square root of the eigenvalues, *i.e.* the factors represent the standard deviations.

From Table 8, the first factor accounts for $3744.08/4452.80 = 84\%$ of the variance in the data. Of this, $22.52^2/3744.08 = 13.5\%$ is derived from object or sample A, 15.2% from B, 21.3% from C, 28.5% from D, and 21.5% from E. The total variance associated with object A is accounted for by the five factors. Taking the square of each element in the factor matrix (remember, these are standard deviations) and summing for each object provides the amount of variance contributed by each object.

For sample A,

$$22.52^2 + 14.78^2 + (-0.71)^2 + 0.68^2 + (-0.08)^2 = 726.6 \qquad (41)$$

and for the other samples,

$$
\begin{aligned}
&\text{B,} \quad 23.82^2 + 12.88^2 + 0.68^2 + 0.52^2 + 0.51^2 = 734.3 \\
&\text{C,} \quad 28.24^2 + 3.20^2 + 0.32^2 + 0.48^2 + 0.77^2 = 808.6 \\
&\text{D,} \quad 32.64^2 + 9.51^2 + 1.02^2 + 0.50^2 + 0.15^2 = 1157.5 \\
&\text{E,} \quad 28.40^2 + 14.78^2 + 0.85^2 + 0.20^2 + 0.45^2 = 1025.8 \qquad (42)
\end{aligned}
$$

These values are identical to the diagonal elements of the variance–covariance matrix from the original data, Table 8, and represent the variance of each object. With all five factors, the total variance is accounted for. If we use fewer factors to explain the data, and this is after all the point of performing a factor analysis, then these totals will be less than 100%. Using just the first two factors, for example, then

$$
\begin{aligned}
&\text{A,} \quad 22.52^2 + 14.78^2 = 725.6, \ 725.6/726.6 = 0.999 \\
&\text{B,} \quad 23.82^2 + 12.88^2 = 733.3, \ 733.3/734.3 = 0.999 \\
&\text{C,} \quad 28.24^2 + 3.20^2 = 807.7, \ 807.7/808.6 = 0.999 \\
&\text{D,} \quad 32.64^2 + 9.51^2 = 1155.8, \ 1155.8/1157.5 = 0.998 \\
&\text{E,} \quad 28.40^2 + 14.78^2 = 1025.0, \ 1025.0/1025.8 = 0.999 \qquad (43)
\end{aligned}
$$

The final values listed in Equation 43 represent the fraction of each object's variance explained by the two factors. They are referred to as the *communality values*, denoted h^2, and they depend on the number of factors used. As the number of factors retained increases, then the communalities tend to unity. The remaining $(1 - h^2)$ fraction of the variance for each sample is considered as being associated with its unique variance and is attributable to noise.

Returning to our mass spectra, having calculated the eigenvalues, eigenvectors, and factor loadings, we must decide how many factors need be retained in our model. In the absence of noise in the measurements, the eigenvalues above the number necessary to describe the data are zero. In practice, of course, noise will always be present. However, as we can see with our mass spectra data a large relative decrease in the magnitude of the eigenvalues occurs after two values, so we can assume here that $p = 2$. Hopke[14] provides an account of several objective functions to assess the correct number of factors.

Having reduced the dimensionality of the data by electing to retain two factors, we can proceed with our analysis and attempt to interpret them. Examination of the columns of loadings for the first two factors in the factor matrix, Table 9, shows that some values are negative. The physical significance of these loadings or coefficients is not immediately apparent. The loadings for these two factors are illustrated graphically in Figure 20(a). The location of the orthogonal vectors in the two-factor space has been constrained by the three remaining but unused factors. If these three factors are not to be used then we can rotate the first two factors in the sample space and possibly find a better position for them; a position which will provide for a more meaningful interpretation of the data. Of the several factor rotation schemes routinely used in factor analysis, that referred to as the *varimax* technique is most commonly used. Varimax rotation moves each factor axis to a new, but still orthogonal, position so that the loading projections are near either the origin or the extremities of these new axes.[8] The rotation is rigid, to retain orthogonality between factors, and is undertaken using an iterative algorithm.

Using the varimax rotation method, the rotated factor loadings for the first two factors from the mass spectra data are given in Table 10 and are plotted in Figure 20(b). The relative position of the objects to each other has remained unchanged, but all loadings are now positive. In fact, all loadings are present in the first quadrant of the diagram and in an order we can recognize as corresponding to the mixtures' compositional analysis. Sample A is predominantly cyclohexane (90%) and E hexane (90%). Examination of Figure 20(b) indicates how we could identify the nature of the two components if they were unknown, as would be the case with a real set of samples of course. Presumably, if the mass spectra of the two pure components were present in, or added to, the original data matrix, then the loadings associated with these samples would align themselves closely with the pure factors. Such, in fact, is the case. Table 11 provides the normalized mass spectra of cyclohexane and hexane, and Table 12 gives the varimax rotated factor loadings for the now 7×7 variance–covariance matrix from the data now containing the mixtures and the pure components. The loadings of the first two factors for the seven samples are illustrated in Figure 21. As expected the single-components spectra are closely aligned with the axes of the derived and rotated factors.

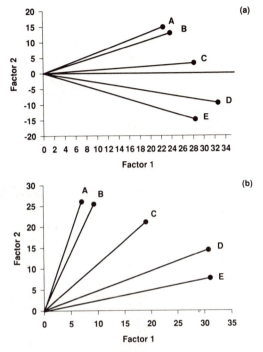

Figure 20 *The original factor loadings obtained from the MS data (a) and the rotated factor loadings, following varimax rotation, with only two factors retained in the model (b)*

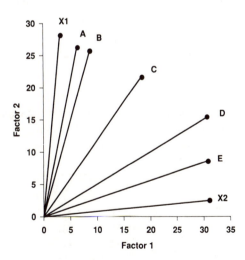

Figure 21 *The rotated factor loadings for the MS data including the pure component spectra, X1 (cyclohexane) and X2 (hexane)*

Table 10 *The rotated factor loading matrix, retaining only the first two factors and using varimax*

	F(I)	F(II)
A	7.16	25.97
B	9.39	25.40
C	19.10	21.04
D	30.80	14.40
E	31.09	7.64

Table 11 *Normalized mass spectral data for cyclohexane and hexane*

m/z	Cyclohexane	Hexane
27	13.3	22.9
28	15.5	16.4
29	9.6	41.8
39	18.5	13.1
40	5.2	3.3
41	52.6	72.1
42	25.9	38.5
43	16.3	70.5
44	1.5	2.5
54	5.9	0.8
55	34.1	7.4
56	100.0	55.7
57	8.9	100.0
69	28.1	1.6
83	5.9	0.8
84	79.3	0.8
85	6.7	0.8
86	0.7	22.9

Table 12 *The rotated factor loadings for the complete MS data set, including the two pure components, cyclohexane (X1) and hexane (X2)*

	F(I)	F(II)
A	3.23	28.05
B	6.45	26.15
C	8.71	25.63
D	18.51	21.57
E	30.69	15.36
X1	30.85	8.50
X2	31.12	2.45

Varimax rotation is a commonly used and widely available factor rotation technique, but other methods have been proposed for interpreting factors from analytical chemistry data. We could rotate the axes in order that they align directly with factors from expected components. These axes, referred to as *test vectors*, would be physically significant in terms of interpretation and the rotation procedure is referred to as *target transformation*. Target transformation factor analysis has proved to be a valuable technique in chemometrics.[21] The number of components in mixture spectra can be identified and the rotated factor loadings in terms of test data relating to standard, known spectra, can be interpreted.

In this chapter we have been able to discuss only some of the more common and basic methods of feature selection and extraction. This area is a major subject of active research in chemometrics. The effectiveness of subsequent data processing and interpretation is largely governed by how well our analytical data have been summarized by these methods. The interested reader is encouraged to study the many specialist texts and journals available to appreciate the wide breadth of study associated with this subject.

[21] P.K. Hopke, D.J. Alpert, and B.A. Roscoe, *Comput. Chem.*, 1983, **7**, 149.

Pattern Recognition I – Unsupervised Analysis

1 Introduction

It is an inherent human trait that, presented with a collection of objects, we will attempt to classify them and organize them into groups according to some observed or perceived similarity. Whether it is with childhood toys and sorting blocks by shape or into coloured sets, or with hobbies devoted to collecting, we obtain satisfaction from classifying things. This characteristic is no less evident in science. Indeed, without the ability to classify and group objects, data, and ideas, the vast wealth of scientific information would be little more than a single random set of data and be of little practical value or use. There are simply too many objects or events encountered in our daily routine to be able to consider each as an individual and discrete entity.

Instead, it is common to refer an observation or measure to some previously catalogued, similar example. The organization in the Periodic Table, for example, allows us to study group chemistry with deviations from general behaviour for any element to be recorded as required. In a similar manner, much organic chemistry can be catalogued in terms of the chemistry of generic functional groups. In infrared spectroscopy, the concept of correlation between spectra and molecular structures is exploited to provide the basis for spectral interpretation; in general each functional group exhibits well defined regions of absorption.

Although the human brain is excellent at recognizing and classifying patterns and shapes, it performs less well if an object is represented by a numerical list of attributes, and much analytical data is acquired and presented in such a form. Consider the data shown in Table 1, obtained from an analysis of a series of alloys. This is only a relatively small data set but it may not be immediately apparent that these samples can be organized into well defined groups defining the type or class of alloy according to their composition. The data from Table 1 are expressed diagrammatically in Figure 1. Although we may guess that there are two similar groups based on the Ni, Cr, and Mn content, the picture suffers from the presence of extraneous data. The situation would be more complex

92

Table 1 *Concentration, expressed as* mg kg^{-1}, *of trace metals in six alloy samples*

	Ni	Cr	Mn	V	Co
A	6.1	1.1	1.2	5.2	4.0
B	1.6	3.8	4.1	4.0	3.9
C	1.2	3.1	2.9	3.6	6.4
D	5.1	1.5	1.6	4.2	3.2
E	4.8	1.8	1.2	3.7	3.0
F	2.1	3.4	4.4	4.3	4.1

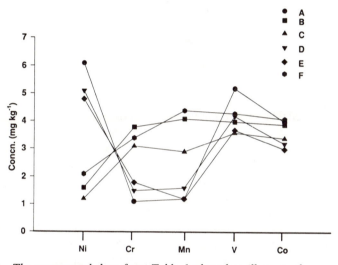

Figure 1 *The trace metal data from Table 1 plotted to illustrate the presence of two groups*

still if more objects were analysed or more variables were measured. As modern analytical techniques are able to generate large quantities of qualitative and quantitative data, it is necessary to seek and apply formal methods which can serve to highlight similarities and differences between samples. The general problem is one of *classification* and the contents of this chapter are concerned with addressing the following, broadly stated task. Given a number of objects or samples, each described by a set of measured values, we are to derive a formal mathematical scheme for grouping the objects into classes such that objects within a class are similar, and different from those in other classes. The number of classes and the class characteristics are not known *a priori* but are to be determined from the analysis.[1]

It is the last statement in the challenge facing us that distinguishes the

---

[1] B. Everitt, 'Cluster Analysis', 2nd Edn, Heinemann Educational, London, UK, 1980.

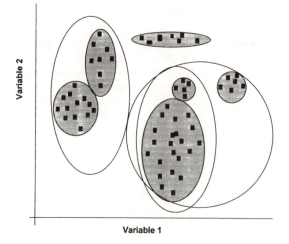

Variable 2

Variable 1

Figure 2   *What constitutes a cluster and its boundary will depend on interpretation as well as the clustering algorithm employed*

techniques studied here from supervised pattern recognition schemes to be examined in Chapter 5. In supervised pattern recognition, a training set is identified with which the parent class or group of each sample is known, and this information is used to develop a suitable discriminant function with which new, unclassified samples can be examined and assigned to one of the parent classes. In the case of *unsupervised pattern recognition*, often referred to as *cluster analysis* or *numerical taxonomy*, no class knowledge is available and no assumptions need be made regarding the class to which a sample may belong. Cluster analysis is a powerful investigative tool, which can aid in determining and identifying underlying patterns and structure in apparently meaningless tables of data. Its use in analytical science is widespread and increasing. Some of the varied areas of its application are model fitting and hypothesis testing, data exploration, and data reduction.[2,3]

The general scheme, or algorithm, followed in order to perform unsupervised pattern recognition and undertake cluster analysis, proceeds in the following manner. The data set comprising the original, or suitably processed, analytical data characterizing our samples is first converted into some corresponding set of similarity, or dissimilarity, measures between each sample. The subsequent aim is to ensure that similar objects are clustered together with minimal separation between objects in a class or cluster, whilst maximizing the separation between different clusters. It is the concept of *similarity* between objects that provides the richness and variety of the wide range of techniques available for cluster analysis. To appreciate this concept, it is worth considering what may constitute a cluster. In Figure 2, two variate measures on a set of 50

[2] J. Chapman, 'Computers in Mass Spectrometry', Academic Press, London, UK, 1978.
[3] H.L.C. Meuzelaar and T.L. Isenhour, 'Computer Enhanced Analytical Spectroscopy', Plenum Press, New York, USA, 1987.

samples are represented in a simple scatter plot. It is evident from visual inspection that there are many ways of dividing the pattern space and producing clusters or groups of objects. There is no single 'correct' result, and the success of any clustering method depends largely on what is being sought, and the intended subsequent use of the information. Clusters may be defined intuitively and their structure and contents will depend on the nature of the problem. The presence of a cluster does not readily admit precise mathematical definition.

2 Choice of Variable

In essence, what all clustering algorithms aim to achieve is to group together similar, neighbouring points into clusters in the n-dimensional space defined by the n-variate measures on the objects. As with supervised pattern recognition (see Chapter 5), and other chemometric techniques, the selection of variables and their pre-processing can greatly influence the outcome. It is worth repeating that cluster analysis is an exploratory, investigative technique and a data set should be examined using several different methods in order to obtain a more complete picture of the information contained in the data.

The initial choice of the measurements made and used to describe each object constitute the frame of reference within which the clusters are to be established. This choice will reflect an analyst's judgement of relevance for the purpose of classification, based usually on prior experience. In most cases the number of variables is determined empirically and often tends to exceed the minimum required to achieve successful classification. Although this situation may guarantee satisfactory classification, the use of an excessive number of variables can severely effect computation time and a method's efficiency. Applying some preprocessing transformation to the data is often worthwhile. Standardization of the raw data can be undertaken, and is particularly valuable when different types of variable are measured. But it should be borne in mind that standardization can have the effect of reducing or eliminating the very differences between objects which are required for classification. Another technique worth considering is to perform a principal components analysis on the original data, to produce a set of new, statistically independent variables. Cluster analysis can then be performed on the first few principal components describing the majority of the samples' variance.

Finally, having performed a cluster analysis, statistical tests can be employed to assess the contribution of each variable to the clustering process. Variables found to contribute little may be omitted and the cluster analysis repeated.

3 Measures between Objects

In general, clustering procedures begin with the calculation of a matrix of similarities or dissimilarities between the objects.[1] The output of the clustering process, in terms of both the number of discrete clusters observed and the cluster membership, may depend on the similarity metric used.

Similarity and distance between objects are complementary concepts for which there is no single formal definition. In practice, distance as a measure of dissimilarity is a much more clearly defined quantity and is more extensively used in cluster analysis.

Similarity Measures

Similarity or *association coefficients* have long been associated with cluster analysis, and it is perhaps not surprising that the most commonly used is the correlation coefficient. Other similarity measures are rarely employed. Most are poorly defined and not amenable to mathematical analysis, and none have received much attention in the analytical literature. The calculation of correlation coefficients is described in Chapter One, and Table 2(a) provides the full symmetric matrix of these coefficients of similarity for the alloy data from Table 1. With such a small data set, a cluster analysis can be performed manually to illustrate the stages involved in the process. The first step is to find the mutually largest correlation in the matrix to form centres for the clusters. The highest correlation in each column of Table 2(a) is shown in boldface type. Objects A and D form mutual highly correlated pairs, as do objects B and C. Note that although object E is most highly correlated with D, they are not considered as forming a pair as D most resembles A rather than E. Similarly, object F is not paired with B, as B is more similar to C.

Table 2   *The matrix of correlations between objects from Table 1, (a). Samples A and D, B and C form new objects and a new correlation matrix can be calculated, (b). Sample E then joins AD and F joins BC to provide the final step and apparent correlation matrix, (c)*

(a)		A	B	C	D	E	F
	A	1	− 0.62	− 0.39	**0.99**	0.97	− 0.44
	B	− 0.62	1	**0.95**	− 0.68	− 0.74	**0.94**
	C	− 0.39	**0.95**	1	− 0.48	− 0.51	0.89
	D	**0.99**	− 0.68	− 0.48	1	**0.98**	− 0.51
	E	0.97	− 0.74	− 0.51	0.98	1	− 0.61
	F	− 0.44	0.94	0.89	− 0.51	− 0.61	1

(b)		AD	BC	E	F
	AD	1	− 0.54	**0.97**	− 0.47
	BC	− 0.54	1	− 0.62	**0.91**
	E	**0.97**	− 0.62	1	− 0.61
	F	− 0.47	**0.91**	− 0.61	1

(c)		ADE	BCF
	ADE	1	− **0.56**
	BCF	− **0.56**	1

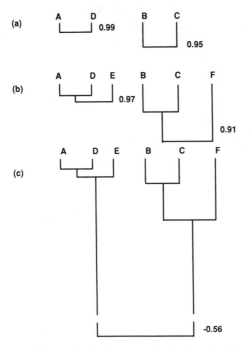

Figure 3 *The stages of hierarchical clustering shown graphically as dendrograms. The steps (a)–(c) correspond to the connections calculated in Table 2*

The resemblance between the mutual pairs is indicated in the diagram shown in Figure 3, which links A to D and B to C by a horizontal line drawn from the vertical axis at points representing their respective correlation coefficients.

At the next stage, objects A and D, and B and C, are considered to comprise new, distinct objects with associative properties and are similar to the other objects according to their average individual values. Table 2(b) shows the newly calculated correlation matrix. Clusters AD and BC have a correlation coefficient calculated from the sum of the correlations of A to B, D to B, A to C, and D to C, divided by four. The correlation between AD and E is the average of the original A to E and D to E correlations. The clustering procedure is now repeated, and object E joins cluster AD and object F joins BC, Figure 3(b). The process is continued until all clusters are joined and the final similarity matrix is produced as in Table 2(c) with the resultant diagram, a *dendrogram*, shown in Figure 3(c). That two groups, ADE and BCF, may be present in the original data is demonstrated.

From this extremely simple example, the basic steps involved in cluster analysis and the value of the technique in classification are evident. The final dendrogram, Figure 3(c), clearly illustrates the similarity between the different samples. The original raw tabulated data have been reduced to a pictorial form which simplifies and demonstrates the structure within the data. It is pertinent to ask, however, what information has been lost in producing the diagram and

Table 3 *The matrix of apparent correlations between the six alloy samples, derived from the dendrogram of Figure 3 and Table 2(b) and (c)*

(a)		A	B	C	D	E	F
	A	1	− 0.56	− 0.56	0.99	0.97	− 0.56
	B	− 0.56	1	0.95	− 0.56	− 0.56	0.91
	C	− 0.56	− 0.95	1	− 0.56	− 0.56	0.91
	D	0.99	− 0.56	− 0.56	1	0.97	− 0.56
	E	0.97	− 0.56	− 0.56	0.97	1	− 0.56
	F	− 0.56	0.91	0.91	− 0.56	− 0.56	1

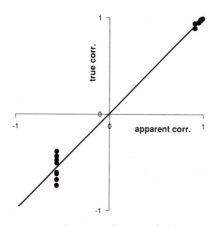

Figure 4 *True vs. apparent correlations indicating the distortion achieved by averaging correlation values to produce the dendrogram*

to what extent does the graph accurately represent our original data. From the dendrogram and Table 2(b) the apparent correlation between sample B and sample F is 0.91, rather than the true value of 0.94 from the calculated similarity matrix. This error arose owing to the averaging process in treating the BC pair as a single entity, and the degree of distortion increases as successive levels of clusters are averaged together. Table 3 is the matrix of apparent correlations between objects as obtained from the dendrogram. These apparent correlations are sometimes referred to as *cophenetic values*, and if these are plotted against actual correlations, Figure 4, then a visual impression is obtained of the distortion in the dendrogram. A numerical measure of the similarity between the values can be calculated by computing the linear correlation between the two sets. If there is no distortion, then the plot would form a straight line and the correlation would be 1. In our example this correlation, $r = 0.99$. Although such a high value for r may indicate a strong linear relationship, as Figure 4 shows there is considerable difference between the real and apparent correlations.

Distance Measures

The correlation coefficient is too limiting in its definition to be of value in many applications of cluster analysis. It is a measure only of colinearity between variates and takes no account of non-linear relationships or the absolute magnitude of variates. Instead, distance measures which can be defined mathematically are more commonly encountered in cluster analysis. Of course, it is always possible at the end of a clustering process to substitute distance with reverse similarity; the greater the distance between objects the less their similarity.

An object is characterized by a set of measures, and it may be represented as a point in multidimensional space defined by the axes, each of which corresponds to a variate. In Figure 5, a data matrix X defines measures of two variables on two objects A and B. Object A is characterized by the pattern vector, $a = x_{11}, x_{12}$, and B by the pattern vector, $b = x_{21}, x_{22}$. Using a distance measure, objects or points closest together are assigned to the same cluster. Numerous distance metrics have been proposed and applied in the scientific literature.

For a function to be useful as a distance metric between objects then the following basic rules apply (for objects A and B):

(a) $d_{AB} \geqslant 0$, the distance between all pairs of measurements for objects A and B must be non-negative,

(b) $d_{AB} = d_{BA}$, the distance measure is symmetrical and can only be zero when A = B.

(c) $d_{AC} + d_{BC} \geqslant d_{AB}$, the distance is commutative for all pairs of points. This statement corresponds to the familiar triangular inequality of Euclidean geometry.

The most commonly referenced distance metric is the *Euclidean distance*, defined by

$$d_{AB} = \left[\sum_j (x_{1_j} - x_{2_j})^2 \right]^{1/2} \tag{1}$$

where $x_{i,j}$ is the value of the j'th variable measured on the i'th object. This equation can be expressed in vector notation as

$$d_{AB} = \left[\sum_j (a_j - b_j)^2 \right]^{1/2} \tag{2}$$

or

$$d_{AB} = [(a - b)^{\mathrm{T}} \cdot (a - b)]^{1/2} \tag{3}$$

In general, most distance metrics conform to the general Minkowski equation,

$$d_{AB} = \left[\sum_j (x_{1_j} - x_{2_j})^m \right]^{1/m} \tag{4}$$

When $m = 1$, Equation (4) defines the *city-block* metric, and if $m = 2$ then the Euclidean distance is defined. Figure 5 illustrates these measures on two-dimensional data.

If the variables have been measured in different units, then it may be necessary to scale the data to make the values comparable.[4-7] An equivalent procedure is to compute a weighted Euclidean distance,

$$d_{AB} = \left[w_j \sum_j (a_j - b_j)^2 \right]^{1/2} \tag{5}$$

or

$$d_{AB} = [(a - b)^T \cdot W(a - b)]^{1/2} \tag{6}$$

W is a symmetric matrix, and in the most simple case W is a diagonal matrix, the diagonal elements of which, w_{ii}, are the weights or coefficients applied to the

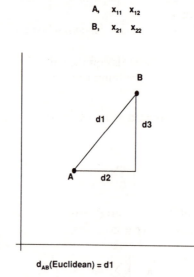

d_{AB}(Euclidean) = d1

d_{AB}(city-block) = d2 + d3

Figure 5   *The Euclidean and city-block metrics for two-dimensional data*

[4] J. Hartigan, 'Clustering Algorithms', J. Wiley and Sons, New York, USA, 1975.
[5] A.A. Afifi and V. Clark, 'Computer Aided Multivariate Analysis', Lifetime Learning, California, USA, 1984.
[6] 'Classification and Clustering', ed. J. Van Ryzin, Academic Press, New York, USA, 1971.
[7] D.J. Hand, 'Discrimination and Classification', J. Wiley and Sons, Chichester, UK, 1981.

vectors corresponding to the variables in the data matrix. Weighting variables is largely subjective and may be based on *a priori* knowledge regarding the data, such as measurement error or equivalent variance of variables. If, for example, weights are chosen to be inversely proportional to measurement variance, then variates with greater precision are weighted more heavily. However, such variates may contribute little to an effective clustering process.

One weighted distance measure which does occur frequently in the scientific literature is the *Mahalanobis distance*,[4,5]

$$d_{AB} = [(a - b)^T \cdot Cov^{-1}(a - b)]^{1/2} \qquad (7)$$

where *Cov* is the full variance–covariance matrtix for the original data. The Mahalanobis distance is invariant under any linear transformation of the original variables. If several variables are highly correlated, this type of weighting scheme down-weights their individual contributions. It should be used with care, however. In cluster analysis, use of the Mahalanobis distance may produce even worse results than equating the variance of each variable and may serve only to decrease the clarity of the clusters.[4]

Before proceeding with a more detailed examination of clustering techniques, we can now compare correlation and distance metrics as suitable measures of similarity for cluster analysis. A simple example serves to illustrate the main points. In Table 4, three objects (A, B, and C) are characterized by five variates. The correlation matrix and Euclidean distance matrix are given in Tables 5 and

Table 4 *Three samples, or objects, characterized by five measures, $x_1 \ldots x_5$*

	x_1	x_2	x_3	x_4	x_5
A	2.1	5.2	3.1	4.1	2.1
B	2.5	4.0	4.0	4.6	3.5
C	5.1	9.2	7.1	7.0	5.0

Table 5 *The correlations matrix of (a), of data from Table 4 and clustering objects with highest mutual correlation, (b)*

(a)

	A	B	C
A	1	0.69	0.96
B	0.69	1	0.63
C	0.96	0.63	1

and the first cluster is AC

(b)

	AC	B
AC	1	0.66
B	0.66	1

Table 6   *The Euclidean distance matrix* (a) *of data from Table 4 and clustering objects with the smallest inter-object distance,* (b)

(a)

	A	B	C
A	0	**2.15**	7.60
B	**2.15**	0	**7.17**
C	7.60	7.17	0

and the first cluster is AB

(b)

	AB	C
AB	0	**7.38**
B	**7.38**	0

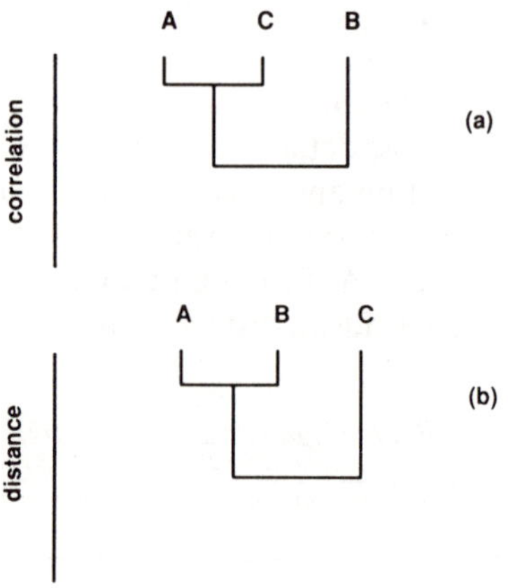

Figure 6   *Dendrograms for the three-object data set from Table 4, clustered according to correlation* (a) *and distance* (b)

6 respectively, and, as before, manual clustering can be undertaken to display the similarity between objects. Using the correlation coefficient, objects A and C form a mutually highly similar pair and may be joined to form a new object AC, with a correlation to object B formed by averaging the A to B, C to B correlations. The resultant dendrogram is shown in Figure 6(a). If the Euclidean distance matrix is used as the measure of similarity, then objects A and B are the most similar as they have the mutually lowest distance separating them. The dendrogram using Euclidean distance is illustrated in Figure 6(b).

Different results may be obtained using different measures. The explanation can be appreciated by considering the original data plotted as in Figure 7. If the variables $x_1, x_2 \ldots x_5$, represent trace elements in, say, water samples and the measures their individual concentrations, then samples A and B would form a group with the differences possibly due to natural variation between samples or experimental error. Sample C could come from a different source. The distance

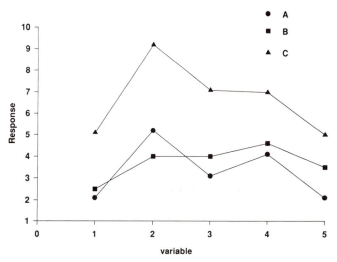

Figure 7 *The three-object from Table 4*

metric in this case provides a suitable clustering measure. On the other hand, if x_1, x_2 . . . x_5 denoted, say, wavelengths and the response values a measure of absorption or emission at these wavelengths, then a different explanation may be sought. It is clear from Figure 7 that if the data represent spectra, then A and C are similar, differing only in scale or concentration, whereas spectrum B has a different profile. Hence, correlation provides a suitable measure of similarity. If, as spectra, the data had been normalized to the most intense response, then A and C would have been closer and the distance metric more meaningful.

In summary, therefore, the first stage in cluster analysis is to compute the matrix of selected distance measures between objects. As the entire clustering process may depend on the choice of distance it is recommended that results using different functions are compared.

4 Clustering Techniques

By grouping similar objects, clusters are themselves representative of those objects and form a distinct group according to some empirical rule. It is implicit in producing clusters that such a group can be represented further by a *typical element* of the cluster. This single element may be a genuine member of the cluster or a hypothetical point, for example an average of the contents' characteristics in multidimensional space. One common method of identifying a cluster's typical element is to substitute the mean values for the variates describing the objects in the cluster. The between-cluster distance can then be defined as the Euclidean distance, or other metric, between these means. Other measures not using the group means are available. The *nearest-neighbour distance* defines the distance between two closest members from different groups. The *furthest-neighbour distance* on the other hand is that between the most remote pair of objects in two groups. A further inter-group measure is obtained by taking the average of all the inter-element measures between elements in different groups. As well as defining the inter-group separation between clusters, each of these measures provides the basis for a clustering technique, defining the method by which clusters are constructed or divided.

In relatively simple cases, in which only two or three variables are measured for each sample, the data can usually be examined visually and any clustering identified by eye. As the number of variates increases, however, this is rarely possible and many scatter plots, between all possible pairs of variates, would need to be produced in order to identify major clusters, and even then clusters could be missed. To address this problem, many numerical clustering techniques have been developed, and the techniques themselves have been classified. For our purposes the methods considered belong to one of the following types.

(a) Hierarchical techniques in which the elements or objects are clustered to form new representative objects, with the process being repeated at different levels to produce a tree structure, the dendrogram.

(b) Methods employing optimization of the partitioning between clusters using some type of iterative algorithm, until some predefined minimum change in the groups is produced.

(c) Fuzzy cluster analysis in which objects are assigned a membership function indicating their degree of belonging to a particular group or set.[8,9]

In order to demonstrate the calculations and results associated with the different methods, the small set of bivariate data in Table 7 will be used. These data comprise 12 objects in two-dimensional space, Figure 8, and the positions

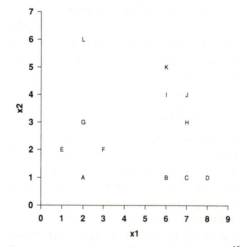

Figure 8   *The bivariate data from Table 7(a)*[10]

[8] A. Kandel, 'Fuzzy Mathematical Techniques with Applications', Addison-Wesley, Massachusetts, USA, 1986.
[9] J.C. Bezdek, 'Pattern Recognition with Fuzzy Objective Function Algorithms', Plenum Press, New York, USA, 1981.
[10] J. Zupan, 'Clustering of Large Data Sets', J. Wiley and Sons, Chichester, UK, 1982.

Table 7 *A simple bivariate data set for cluster analysis (a), from Zupan,[10] and the corresponding Euclidean distance matrix, (b)*

(a)		A	B	C	D	E	F	G	H	I	J	K	L
	X_1	2	6	7	8	1	3	2	7	6	7	6	2
	X_2	1	1	1	1	2	2	3	3	4	4	5	6

(b)	A	B	C	D	E	F	G	H	I	J	K	L
A	0	4.0	5.0	6.0	**1.4**	**1.4**	2.0	5.4	5.0	5.8	5.7	5.0
B	4.0	0	1.0	2.0	5.1	3.2	4.5	2.2	3.0	3.2	4.0	6.4
C	5.0	**1.0**	0	**1.0**	6.1	4.1	5.4	2.0	3.2	3.0	4.1	7.1
D	6.0	2.0	**1.0**	0	7.1	5.1	6.3	2.2	3.6	3.2	4.5	7.8
E	**1.4**	5.1	6.1	7.1	0	2.0	**1.4**	6.1	5.4	6.3	5.8	4.1
F	**1.4**	3.2	4.1	5.1	2.0	0	**1.4**	4.1	3.6	4.5	4.2	4.1
G	2.0	4.5	5.4	6.3	**1.4**	**1.4**	0	5.0	4.1	5.1	4.5	**3.0**
H	5.4	2.2	2.0	2.2	6.1	4.1	5.0	0	1.4	**1.0**	2.2	5.8
I	5.0	3.0	3.2	3.6	5.4	3.6	4.1	1.4	0	**1.0**	**1.0**	4.5
J	5.8	3.2	3.0	3.2	6.3	4.5	5.1	**1.0**	**1.0**	0	1.4	5.4
K	5.7	4.0	4.1	4.5	5.8	4.2	4.5	2.2	**1.0**	1.4	0	4.1
L	5.0	6.4	7.1	7.8	4.1	4.1	3.0	5.8	4.5	5.4	4.1	0

of the points are representative of different shaped clusters, the single point (L), the extended group (B,C,D), the symmetrical group (A,E,F,G), and the asymmetrical cluster (H,I,J,K).[10]

Hierarchical Techniques

When employing hierarchical clustering techniques, the original data are separated into a few general classes, each of which is further subdivided into still smaller groups until finally the individual objects themselves remain. Such methods may be agglomerative or divisive. By agglomerative clustering, small groups, starting with individual samples, are fused to produce larger groups as in the examples studied previously. In contrast, divisive clustering starts with a single cluster, containing all samples, which is successively divided into smaller partitions. Hierarchical techniques are very popular, not least because their application leads to the production of a dendrogram which can provide a two-dimensional pictorial representation of the clustering process and the results. Agglomerative hierarchical clustering is very common and we will proceed with details of its application.

Agglomerative methods begin with the computation of a similarity or distance matrix between the objects, and result in a dendrogram illustrating the succesive fusion of objects and groups until the stage is reached when all objects are fused into one large set. Agglomerative methods are the most common hierarchical schemes found in scientific literature. The entire process involved in undertaking agglomerative clustering using distance measures can be summarized by a four-step algorithm.

Step 1. Calculation of the between-object distance matrix.

Step 2. Find the smallest elements in the distance matrix and join the corresponding objects into a single cluster.

Step 3. Calculate a new distance matrix, taking into account that clusters produced in the second step will have formed new objects and taken the place of original data points.

Step 4. Return to Step 2 or stop if the final two clusters have been fused into the final, single cluster.

The wide range of agglomerative methods available differ principally in the implementation of Step 3 and the calculation of the distance between two clusters. The different between-group distance measures can be defined in terms of the general formula

$$d_{k(i,j)} = \alpha_i d_{k,i} + \alpha_j d_{k,j} + \beta d_{i,j} + \gamma |d_{k,i} - d_{k,j}| \tag{8}$$

where $d_{i,j}$ is the distance between objects i and j and $d_{k(i,j)}$ is the distance between group k and a new group (i,j) formed by the fusion of groups i and j. The values of coefficients α_i, α_j, β, and γ are chosen to select the specific between-group metric to be used. Table 8 lists the more common metrics and the corresponding values for α_i, α_j, β, and γ.

The use of Equation (8) makes it a simple matter for standard computer software packages to offer a choice of distance measures to be investigated by selecting the appropriate values of the coefficients.

Table 8 *The common distance measures used in cluster analysis*

Metric	Coefficients			
	a_i	a_j	b	d
Nearest neighbour (single linkage)	0.5	0.5	0	-0.5
Furthest neighbour (complete linkage)	0.5	0.5	0	0.5
Centroid	$\dfrac{n_i}{n_i + n_j}$	$\dfrac{n_j}{n_i + n_j}$	$-a_i . a_j$	
Median	0.5	0.5	-0.25	0
Group average	$\dfrac{n_i}{n_i + n_j}$	$\dfrac{n_j}{n_i + n_j}$	0	0
Ward's method	$\dfrac{n_k + n_i}{n_k + n_j + n_i}$	$\dfrac{n_k + n_j}{n_k + n_i + n_j}$	$\dfrac{-n_k}{n_k + n_i + n_j}$	0

The number of objects in any cluster i is n_i.

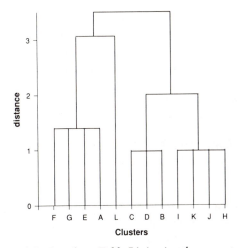

Figure 9 *Dendrogram of the data from Table 7(a) using the nearest-neighbour algorithm*

For the *nearest-neighbour* method of producing clusters, Equation (8) reduces to

$$d_{k(i,j)} = 0.5d_{k,i} + 0.5d_{k,j} - 0.5|d_{k,i} - d_{k,j}| \tag{9}$$

From the 12 × 12 distance matrix, Table 7(b), objects B and C form a new, combined object and the distance from BC to each original object is calculated according to Equation (9). Thus, for A to BC,

$$\begin{aligned} d_{A(B,C)} &= 0.5d_{AB} + 0.5d_{AC} - 0.5|d_{AB} - d_{AC}| \\ &= 0.5(4) + 0.5(5) - 0.5(1) \\ &= 4 \end{aligned} \tag{10}$$

In fact, for the nearest-neighbour algorithm, Equation (9) can be rewritten as

$$d_{k(i,j)} = \min(d_{k,i}, d_{k,j}) \tag{11}$$

i.e. the distance between a cluster and an object is the smallest of the distances between the elements in the cluster and the object.

The distance between the new object BC and each remaining original object is calculated, and the procedure repeated with the resulting 11 × 11 distance matrix until a single cluster containing all objects is produced. The resulting dendrogram is illustrated in Figure 9.

The dendrogram for the *furthest-neighbour*, or *complete linkage*, technique is produced in a similar manner. In this case, Equation (8) becomes

$$d_{k(i,j)} = 0.5d_{k,i} + 0.5d_{k,j} + 0.5|d_{k,i} - d_{k,j}| \tag{12}$$

and this implies that

$$d_{k(i,j)} = \max(d_{k,i}, d_{k,j}) \tag{13}$$

i.e. the distance between a cluster and an object is the maximum of the distances between cluster elements and the object.

For example, for group BC to object D, the B to D distance is 2 units and the C to D distance is 1 unit. From Equation (13), therefore $d_{D(BC)} = 2$, or

$$\begin{aligned} d_{D(BC)} &= 0.5d_{DB} + 0.5d_{DC} - 0.5|d_{DB} - d_{DC}| \\ &= 0.5(2) + 0.5(1) + 0.5(1) \\ &= 2 \end{aligned} \tag{14}$$

The complete dendrogram is shown in Figure 10. The nearest-neighbour and furthest-neighbour criteria are the simplest algorithms to implement.

Another procedure, *Ward's method*, is commonly encountered in chemometrics. A centroid point is calculated for all combinations of two clusters and the distance between this point and all other objects calculated. In practice this technique generally favours the production of small clusters.

From Equation (8),

$$\begin{aligned} d_{D(BC)} &= 2d_{DB}/3 + 2d_{DC}/3 - 1d_{BC}/3 \\ &= 2(2)/3 + 2(1)/3 - 1(1)/3 \\ &= 1.67 \end{aligned} \tag{15}$$

The process is repeated between BC and other objects, and the iteration starts again with the new distance matrix until a single cluster is produced. The dendrogram from applying Ward's method is illustrated in Figure 11.

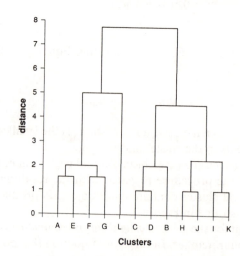

Figure 10 *Dendrogram of the data from Table 7(a) using the furthest-neighbour algorithm*

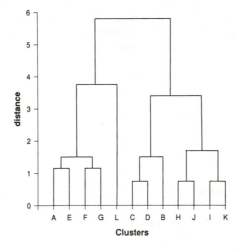

Figure 11 *Dendrogram of the data from Table 7(a) using Ward's method*

The different methods available from applying Equation (8) with the co-efficients from Table 8 each produce their own style of dendrogram with their own merits and disadvantages. Which technique or method is best is largely governed by experience and empirical tests. The construction of the dendrogram invariably induces considerable distortion as discussed, and other, non-hierarchical, methods are generally favoured when large data sets are to be analysed.

The *K*-Means Algorithm

One of the most popular and widely used clustering techniques is the application of the *K*-Means algorithm. It is available with all popular cluster analysis software packages and can be applied to relatively large sets of data. The objective of the method is to partition the m objects, characterized by n variables, into K clusters so that the square of the within-cluster sum of distances is minimized. Being an optimization-based technique, the number of possible solutions cannot be predicted and the best possible partitioning of the objects may not be achieved. In practice, the method finds a local optimum, defined as being a classification in which no movement of an observation from one cluster to another will reduce the within-cluster sum of squares.

Many versions of the algorithm exist, but in most cases the user is expected to supply the number of clusters, K, expected. The algorithm described here is that proposed by Hartigan.[4]

The data matrix is defined by X with elements $x_{i,j}$, $(1 \leq i \leq m, 1 \leq j \leq n)$, where m is the number of objects and n is the number of variables used to characterize the objects. The cluster analysis seeks to find K partitions or clusters, with each object residing in only one of the clusters.

The mean value for each variable j, for all objects in cluster L is denoted by $B_{L,j}$, $(1 \leqslant L \leqslant K)$. The number of objects residing in cluster L is R_L.

The distance, $D_{i,L}$, between the i'th object and the centre or average of each cluster is given by the Euclidean metric,

$$D_{i,L} = [(x_{i,j} - B_{L,j})^2]^{1/2} \tag{16}$$

and the error associated with any partition is

$$\epsilon = \Sigma(D_{iL(i)})^2 \tag{17}$$

where $L(i)$ is the cluster containing the i'th object. Thus ϵ represents the sum of the squares of the distances between object i and the cluster centres.

The algorithm proceeds by moving an object from one cluster to another in order to reduce ϵ, and ends when no movement can reduce ϵ. The steps involved are:

Step 1: Given K clusters and their initial contents, calculate the cluster means $B_{L,j}$ and the initial partition error, ϵ.

Step 2: For the first object, compute the increase in error, $\Delta\epsilon$, obtained by transferring the object from its current cluster, $L(1)$, to every other cluster L, $2 \leqslant L \leqslant K$:

$$\Delta\epsilon = \frac{(R_L)(D_{1,L})^2}{(R_L) + 1} - \frac{(R_{L(1)})(D_{1,L(1)})^2}{(R_{L(1)}) - 1} \tag{18}$$

If this value for $\Delta\epsilon$ is negative, *i.e.* the move would reduce the partition error, transfer the object and adjust the cluster means taking account of their new populations.

Step 3: Repeat Step 2 for every object.

Step 4: If no object has been moved then stop, else return to Step 2.

Applying the algorithm manually to our test data will illustrate its operation.

Using the data from Table 7, it is necessary first to specify the number of clusters into which the objects are to be partitioned. We will use $K = 4$. Before the algorithm is implemented we also need to assign each object to an initial cluster. A number of methods are available, and that used here is to assign object i to cluster $L(i)$ according to

$$L(i) = \mathrm{INT}\left[(K - 1)\left(\frac{\Sigma X_{i,j} - \mathrm{MIN}\,\Sigma X_{i,j}}{\mathrm{MAX}\,\Sigma X_{i,j} - \mathrm{MIN}\,\Sigma X_{i,j}}\right)\right] + 1 \tag{19}$$

where $\sum_j x_{i,j}$ is the sum of all variables for each object, and MIN and MAX denote the minimum and maximum sum values.

For the test data,

Variables	Objects											
	A	B	C	D	E	F	G	H	I	J	K	L
x_1	2	6	7	8	1	3	2	7	6	7	6	2
x_2	1	1	1	1	2	2	3	3	4	4	5	6
$\Sigma X_{i,j}$	3	7	8	9	3	5	5	10	10	11	11	8

$$\text{MAX } \Sigma X_{i,j} = 11 \qquad\qquad \text{MIN } \Sigma X_{i,j} = 3$$

For object A,

$$L(A) = \text{INT}\{(4 - 1)[(3 - 3)/(11 - 3)]\} + 1 = 1 \qquad (20)$$

and similarly for each object, all i,

$$
\begin{array}{llllllllllll}
i = & \text{A} & \text{B} & \text{C} & \text{D} & \text{E} & \text{F} & \text{G} & \text{H} & \text{I} & \text{J} & \text{K} & \text{L} \\
L(i) = & 1 & 2 & 2 & 3 & 1 & 1 & 1 & 3 & 3 & 4 & 4 & 2
\end{array}
$$

Thus, objects A, E, F, G are assigned initially to Cluster 1, B, C, and L to Cluster 2, D, H, and I to Cluster 3, and finally J and K to Cluster 4.

The centres of each of these clusters can now be calculated. For Cluster 1, $L = 1$,

$$B_{1,1} = (2 + 1 + 3 + 2)/4 = 2.00 \qquad (21)$$
$$B_{1,2} = (1 + 2 + 2 + 3)/4 = 2.00 \qquad (22)$$

Similarly for each of the remaining three clusters. The initial clusters and their mean values are therefore,

Cluster	Contents	Cluster means	
		x_1	x_2
1	A E F G	2.00	2.00
2	B C L	5.00	2.67
3	D H I	7.00	2.67
4	J K	6.50	4.50

This initial partitioning is illustrated in Figure 12(a).

By application of Equations (16) and (17), the error associated with this initial partitioning is

$$
\begin{aligned}
\epsilon = &\ (2 - 2)^2 + (1 - 2)^2 + (6 - 5)^2 + (1 - 2.67)^2 + (7 - 5)^2 \\
&+ (1 - 2.67)^2 + (8 - 7)^2 + (1 - 2.67)^2 + (1 - 2)^2 + (2 - 2)^2 \\
&+ (3 - 2)^2 + (2 - 2)^2 + (2 - 2)^2 + (3 - 2)^2 + (7 - 7)^2 \\
&+ (3 - 2.67)^2 + (6 - 7)^2 + (4 - 2.67)^2 + (7 - 6.5)^2 \\
&+ (4 - 4.5)^2 + (6 - 6.5)^2 + (5 - 4.5)^2 + (2 - 5)^2 + (6 - 2.67)^2 \\
= &\ 42.35
\end{aligned} \qquad (23)
$$

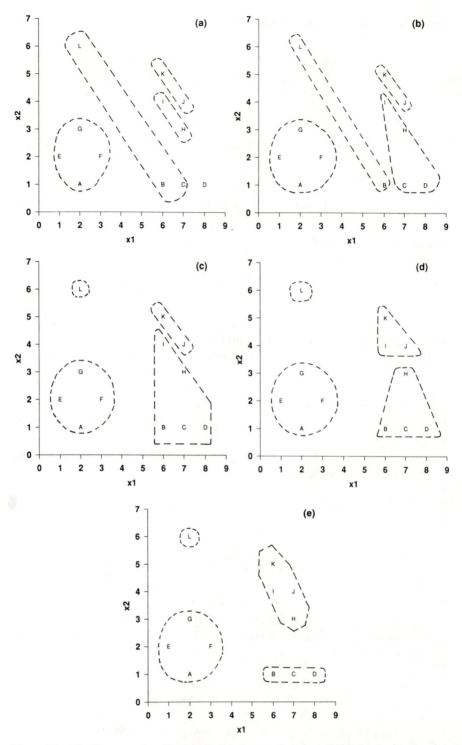

Figure 12 *The K-means algorithm applied to the test data from Table 7(a), showing the initial four partitions* (a) *and subsequent steps,* (b) *to* (d), *until a stable result is achieved* (e)

In order to reduce this error, the algorithm proceeds to examine each object in turn and calculate the effect of transferring an object to a new cluster.

For the first object, A, its squared Euclidean distance to each cluster centre is

$$
\begin{aligned}
D_{A,1}{}^2 &= (2.00 - 2.00)^2 + (1.00 - 2.00)^2 &= 1.00 \\
D_{A,2}{}^2 &= (2.00 - 5.00)^2 + (0.001 - 2.67)^2 &= 11.79 \\
D_{A,3}{}^2 &= (2.00 - 7.00)^2 + (1.00 - 2.67)^2 &= 17.79 \\
D_{A,4}{}^2 &= (2.00 - 6.50)^2 + (1.00 - 4.50)^2 &= 32.50
\end{aligned}
\tag{24}
$$

If we were to transfer object A from Cluster 1 to Cluster 2, then the change in error, from Equation (18), is

$$
\Delta\epsilon = (3)(11.79)/4 - (4)(1)/3 = 7.51 \tag{25}
$$

and to Cluster 3,

$$
\Delta\epsilon = (3)(27.79)/4 - (4)(1)/3 = 19.51 \tag{26}
$$

and to Cluster 4,

$$
\Delta\epsilon = (3)(32.50)/4 - (4)(1)/3 = 20.34 \tag{27}
$$

The $\Delta\epsilon$ values are all positive and each proposed change would serve to increase the partition error, so object A is not moved from Cluster 1. This result can be appreciated by reference to Figure 12(a). Object A is closest to the centre of Cluster 1 and nothing would be gained by assigning it to another cluster.

The algorithm continues by checking each object and calculating $\Delta\epsilon$ for each object with each cluster. For our purpose, visual examination of Figure 12(a) indicates that no change would be expected for object B, but for object C a move is likely as it is closer to the centre of Cluster 3 than Cluster 2.

Moving object C, the third object, to Cluster 1,

$$
D_{C,1}{}^2 = (7.00 - 2.00)^2 + (1.00 - 2.00)^2 = 26.00
$$

and

$$
\Delta\epsilon \quad = (4)(26.00)/5 - (3)(6.79)/2 = 3.82 \tag{28}
$$

for Cluster 2, its current group,

$$
D_{C,2}{}^2 = (7.00 - 5.00)^2 + (1.00 - 2.67)^2 = 6.79 \tag{29}
$$

and to Cluster 3,

$$
D_{C,3}{}^2 = (7.00 - 7.00)^2 + (1.00 - 2.67)^2 = 2.79
$$

and

$$\Delta\epsilon \quad = (3)(2.79)/4 - (3)(6.79)/2 = -14.88 \tag{30}$$

and to Cluster 4,

$$D_{C,4}^2 = (7.00 - 6.5)^2 + (1.00 - 4.50)^2 = 12.50$$

and

$$\Delta\epsilon \quad = (2)(12.50)/3 - (3)(6.79)/2 = -8.64 \tag{31}$$

So, moving object C from Cluster 2 to Cluster 3 decreases ϵ by 14.88, and the new value of ϵ is $(42.35 - 14.88) = 27.47$. The partition is therefore changed. With new clusters and contents we must calculate their new mean values:

Cluster	Contents (1st change)	Cluster means x_1	x_2
1	A E F G	2.00	2.00
2	B L	4.00	3.50
3	C D H I	7.00	2.50
4	J K	6.50	4.50

The new partition, after the first pass through the algorithm, is illustrated in Figure 12(b). On the second run through the algorithm object B will transfer to Cluster 3; it is nearer its mean than Cluster 2.

On the second pass, therefore, the cluster populations and their centres are, Figure 12(c),

Cluster	Contents (2nd change)	Cluster means x_1	x_2
1	A E F G	2.00	2.00
2	L	2.00	6.00
3	B C D H I	6.80	2.00
4	J K	6.50	4.50

On the next pass, object I will move to Cluster 4, Figure 12(d),

Cluster	Contents (3rd change)	Cluster means x_1	x_2
1	A E F G	2.00	2.00
2	L	2.00	6.00
3	B C D H	7.00	1.50
4	I J K	6.33	4.33

On the fourth pass, object H moves from Cluster 3 to Cluster 4, Figure 12(e),

Cluster	Contents (4th change)	Cluster means	
		x_1	x_2
1	A E F G	2.00	2.00
2	L	2.00	6.00
3	B C D	7.00	1.00
4	H I J K	6.50	4.00

The process is repeated once more but this time no movement of any object between clusters gives a better solution in terms of reducing the value of ϵ. So Figure 12(e) represents the best result.

Our initial assumption when applying the K-means algorithm was that four clusters were known to exist. Visual examination of the data suggests that this assumption is reasonable in this case, but other values could be acceptable depending on the model investigated. For $K = 2$ and $K = 3$, the K-means algorithm produces the results illustrated in Figure 13. Although statistical tests have been proposed in order to select the best number of partitions, cluster analysis is not generally considered a statistical technique, and the choice of criteria for best results is at the discretion of the user.

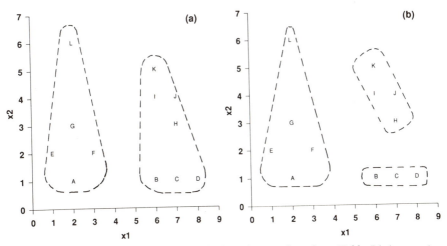

Figure 13 *The K-means algorithm applied to the test data from Table 7(a) assuming there are two clusters (a), and three clusters (b)*

Fuzzy Clustering

The principal aim of performing a cluster analysis is to permit the identification of similar samples according to their measured properties. Hierarchical techniques, as we have seen, achieve this by linking objects according to some formal rule set. The K-means method on the other hand seeks to partition the pattern space containing the objects into an optimal predefined number of

Table 9 *Bivariate data (x_1 and x_2) measured on 15 objects, A ... O*

Object, i	Variable, j		Object, i	Variable, j	
	x_1	x_2		x_1	x_2
A	1	1	I	5	3
B	1	3	J	6	2
C	1	5	K	6	3
D	2	2	L	6	4
E	2	3	M	7	1
F	2	4	N	7	3
G	3	3	O	7	5
H	4	3			

sections. In the process of providing a simplified representation of the data, both schemes can distort the 'true' picture. By linking similar objects and reducing the data to a two-dimensional histogram, hierarchical clustering often severely distorts the similarity value by averaging values or selecting maximum or minimum values. The result of *K*-means clustering is a simple list of clusters, their centres, and their contents. Nothing is said about how well any specific object fits into its chosen cluster, or how close it may be to other clusters. In Figure 12(e), for example, object C is more representative of its parent cluster than, say, object B which may be considered to have some of the characteristics of the first cluster containing objects A, E, F, G.

One clustering method which seeks to not only highlight similar objects but also provide information regarding the relationship of each object to each cluster is that of *fuzzy clustering*.[8,9] Generally referred to in the literature as the *fuzzy c-means* method, to preserve continuity of the symbols used thus far, we will identify the technique as fuzzy *k*-means clustering. The method illustrated here is based on Bezdek's algorithm.[9]

In order to demonstrate the use and application of fuzzy clustering, a simple set of data will be analysed manually. The data in Table 9 represent 15 objects (A ... O) characterized by two variables x_1 and x_2, and these data are plotted in the scatter diagram of Figure 14. It is perhaps not unreasonable to assume that these data represent two classes or clusters. The means of the clusters are well separated but the clusters touch about points G, H, and I. Because the apparent groups are not well separated, the results using conventional cluster analysis schemes can be misleading or ambiguous. With the data from Table 9 and applying the *K*-means algorithm using two different commercially available software packages, the results are as illustrated in Figure 15(a) and 15(b). These results are confusing. Since the data are symmetrical, in the x_2 axis, about $x_2 = 3$, why should points B, E, G, H, I, K, N belong to one cluster rather than the other cluster? Similarly, in Figure 15(b), in the x_1 axis the data are symmetrical about $x_1 = 4$ and there is no reason why object H should belong exclusively to either cluster. The problem arises because of the *crisp* nature of the clustering rule that assigns each object to one specific cluster. This rule is

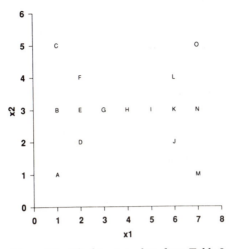

Figure 14 *The bivariate data from Table 9*

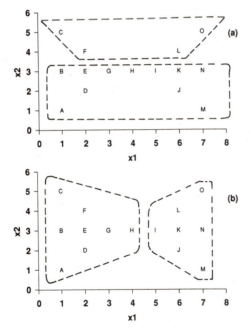

Figure 15 *Clustering resulting from application of two commercial programs of the K-means algorithm* (a) *and* (b), *to the data from Table 9*

relaxed when applying fuzzy clustering and objects are recognized as belonging, to a lesser or greater degree, to every cluster.

The degree or extent to which an object, i, belongs to a specific cluster, k, is referred to as that object's *membership function*, denoted μ_{ki}. Thus, visual inspection of Figure 14 would suggest that for two clusters objects E and K would be close to the cluster centres, *i.e.* $\mu_{1E} \sim 1$ and $\mu_{2K} \sim 1$, and that object H would belong equally to both clusters, *i.e.* $\mu_{1H} = 0.5$ and $\mu_{2H} = 0.5$. This is precisely the result obtained with fuzzy clustering.

As with *K*-means clustering, the fuzzy *k*-means technique is iterative and seeks to minimize the within-cluster sum of squares. Our data matrix is defined by the elements x_{ij} and we seek *K* clusters, not by hard partitioning of the variable space, but by fuzzy partitions, each of which has a cluster centre or prototype value, B_{kj}, $(1 < k < K)$.

The algorithm starts with a pre-selected number of clusters, *K*. In addition, an initial fuzzy partition of the objects is supplied such that there are no empty clusters and the membership functions for an object with respect to each cluster sum to unity,

$$\mu_{1A} + \mu_{2A} + \ldots + \mu_{KA} = 1 \tag{32}$$

Thus, if $\mu_{1E} = 0.8$, then $\mu_{2E} = 0.2$.

The algorithm proceeds by calculating the *K*-weighted means in order to determine cluster centres,

$$B_{kj} = \sum_{i=1}^{M} (\mu_{ki})^2 \cdot x_{ij} \Bigg/ \sum_{i=1}^{M} (\mu_{ki})^2 \tag{33}$$

New fuzzy partitions are then defined by a new set of membership functions given by,

$$\mu_{ki} = \frac{1}{\sum_{j=1}^{N}(x_{ij} - B_{kj})^2} \Bigg/ \sum_{k=1}^{K}\left(\frac{1}{\sum_{j=1}^{N}(x_{ij} - B_{kj})^2}\right) \tag{34}$$

i.e. the ratio of the inverse squared distance of object *i* from the *k*'th cluster centre to the sum of the inverse squared distances of object *i* to all cluster centres.

From this new partitioning, new cluster centres are calculated by applying Equation (33), and the process repeats until the total change in values of the membership functions is less than some preselected value, or a set number of iterations has been achieved.

Application of the algorithm can be demonstrated using the data from Table 9.

With $K = 2$, our first step is to assign membership functions for each object and each cluster. This process can be done in a random fashion, bearing in mind the constraint imposed by Equation (32), or using prior knowledge, *e.g.* the output from crisp clustering methods. With the results from the *K*-means algorithm, Figure 15(b), the membership functions can be assigned as shown in Table 10. Objects A . . . H belong predominately to Cluster 1 and objects I . . . O to Cluster 2.

Table 10 *Initial membership functions, μ_{ki}, for 20 objects assuming two clusters*

i	μ_{1i}	μ_{2i}	i	μ_{1i}	μ_{2i}
A	0.9	0.1	I	0.1	0.9
B	0.9	0.1	J	0.1	0.9
C	0.9	0.1	K	0.1	0.9
D	0.9	0.1	L	0.1	0.9
E	0.9	0.1	M	0.1	0.9
F	0.9	0.1	N	0.1	0.9
G	0.9	0.1	O	0.1	0.9
H	0.9	0.1			

Using this initial fuzzy partition, the initial cluster centres can be calculated according to Equation (33).

$$B_{11} = \sum_{i=1}^{15} (\mu_{1i})^2 . x_{i1} \Big/ \sum_{i=1}^{15} (\mu_{1i})^2$$

$$
\begin{aligned}
&= [(0.9^2)1 + (0.9^2)1 + (0.9^2)1 + (0.9^2)2 + (0.9^2)2 + (0.9^2)2 \\
&\quad + (0.9^2)3 + (0.9^2)4 + (0.1^2)5 + (0.1^2)6 + (0.1^2)6 + (0.1^2)6 \\
&\quad + (0.1^2)7 + (0.1^2)7 + (0.1^2)7]/[8(0.9^2) + 7(0.1^2)] \\
&= 2.04
\end{aligned}
\tag{35}
$$

Similarly for B_{12}, B_{21}, B_{22}, and the centres are

$$B_{11} = 2.04, \qquad B_{12} = 3.00$$
$$B_{21} = 6.20, \qquad B_{22} = 3.00$$

And we can proceed to calculate new membership functions for each object about these centres.

The squared Euclidean distance between object A and the centre of Cluster 1 is, from Equation (1),

$$
\begin{aligned}
d_{AB(1)} &= \sum_{j=1}^{2} (x_{1j} - B_{1j})^2 \\
&= (1 - 2.04)^2 + (1 - 3.00)^2 \\
&= 5.08
\end{aligned}
\tag{37}
$$

and to Cluster 2, $d_{AB(2)} = 31.04$. The new membership functions for object A, from Equation (34), are therefore,

$$\mu_{1A} = \frac{1/5.08}{(1/5.08) + (1/31.04)} = 0.86 \tag{38}$$

and

Table 11 *Final membership functions, μ_{ki}, for 20 objects assuming two clusters*

i	μ_{1i}	μ_{2i}	i	μ_{1i}	μ_{2i}
A	086	0.14	I	0.12	0.88
B	0.97	0.03	J	0.06	0.94
C	0.86	0.14	K	0.01	0.99
D	0.94	0.06	L	0.06	0.94
E	0.99	0.01	M	0.14	0.86
F	0.94	0.06	N	0.03	0.97
G	0.88	0.12	O	0.14	0.86
H	0.05	0.50			

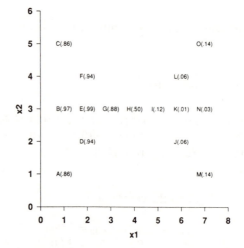

Figure 16 *Results of applying the fuzzy k-means clustering algorithm to the data from Table 9. The values in parenthesis indicate the membership function for each object relative to group A*

$$\mu_{2A} = \frac{1/31.04}{(1/5.08) + (1/31.04)} = 0.14 \qquad (39)$$

The sum ($\mu_{1A} + \mu_{2A}$) is unity, which satisfies Equation (32), and the membership functions for the other objects can be calculated in a similar manner. The process is repeated and after five iterations the total change in the squared μ_{ki} values is less than 10^{-5} and the membership functions are considered stable, Table 11. This result, Figure 16, accurately reflects the symmetric distribution of the data.

The same algorithm that provides the membership functions for the test data can be used to generate values for interpolated and extrapolated data, and a three-dimensional surface plot produced, Figure 16(b). Going one stage further, we can combine μ_{1i} and μ_{2i} to provide the complete membership surface, according to the rule

$$\mu_i = \max(\mu_{1i}, \mu_{2i}) \tag{40}$$

This result is illustrated in Figure 17.

Fuzzy clustering can be applied to the test data examined previously, from Table 7, and the results obtained when two, three, and four clusters are initially specified are provided in Table 12. As with using the K-means algorithm, although various parameters have been proposed to select the best number of clusters, no single criterion is universally accepted. Although the results of fuzzy clustering may appear appealing, we are still left with the need to make some decision as to which cluster an object belongs. This may be achieved by specifying some threshold membership value, α, in order to identify the core of the cluster. Thus, if say $\alpha = 0.5$, then from Figure 16 objects A, B, C, D, E, F, G belong to Cluster 1, I, J, K, L, M, N, O can be assigned to Cluster 2, and object H is an outlier from the two clusters.

Table 12 *Membership function values for the objects from Table 7 assuming two, three, and four clusters in the data*

	Two clusters		Three clusters			Four clusters			
	μ_{1i}	μ_{2i}	μ_{1i}	μ_{2i}	μ_{3i}	μ_{1i}	μ_{2i}	μ_{3i}	μ_{4i}
A	0.09	0.99	0.06	0.06	0.88	0.03	0.91	0.03	0.03
B	0.83	0.17	0.87	0.08	0.05	0.83	0.05	0.09	0.02
C	0.89	0.11	1.0	0.0	0.0	1.0	0.0	0.0	0.0
D	0.89	0.11	0.90	0.07	0.03	0.89	0.02	0.0	0.01
E	0.04	0.96	0.03	0.03	0.94	0.02	0.90	0.02	0.05
F	0.07	0.93	0.06	0.07	0.87	0.05	0.84	0.05	0.05
G	0.01	0.99	0.01	0.02	0.97	0.03	0.82	0.04	0.11
H	0.99	0.01	0.37	0.58	0.05	0.26	0.04	0.67	0.03
I	0.89	0.11	0.01	0.98	0.01	0.02	0.01	0.96	0.01
J	0.94	0.06	0.08	0.90	0.02	0.03	0.01	0.94	0.01
K	0.79	0.21	0.04	0.94	0.02	0.05	0.03	0.86	0.05
L	0.27	0.83	0.14	0.33	0.53	0.0	0.0	0.0	1.0

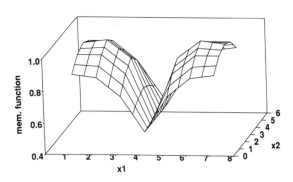

Figure 17 *The two cluster surface plot of data from Table 9 using the fuzzy clustering algorithm*

Cluster analysis is justifiably a popular and common technique for exploratory data analysis. Most commercial multivariate statistical software packages offer several algorithms, along with a wide range of graphical display facilities to aid the user in identifying patterns in data. Having indicated that some pattern and structure may be present in our data, it is often necessary to examine the relative importance of the variables and determine how the clusters may be defined and separated. This is the primary function of supervised pattern recognition and is examined in Chapter 5.

CHAPTER 5

Pattern Recognition II: Supervised Learning

1 Introduction

Generally, the term pattern recognition tends to refer to the ability to assign an object to one of several possible categories, according to the values of some measured parameters. In statistics and chemometrics, however, the term is often used in two specific areas. In Chapter 4, unsupervised pattern recognition, or cluster analysis, was introduced as an exploratory method for data analysis. Given a collection of objects, each of which is described by a set of measures defining its pattern vector, cluster analysis seeks to provide evidence of natural groupings or clusters of the objects in order to allow the presence of patterns in the data to be identified. The number of clusters, their populations, and their interpretation are somewhat subjectively assigned and are not known before the analysis is conducted. Supervised pattern recognition, the subject of this chapter, is very different, and is often referred to in the literature as classification or discriminant analysis. With supervised pattern recognition, the number of parent groups is known in advance and representative samples of each group are available. With this information, the problem facing the analyst is to assign an unclassified object to one of the parent groups. A simple example will serve to make this distinction between unsupervised and supervised pattern recognition clearer.

Suppose we have determined the elemental composition of a large number of mineral samples, and wish to know whether these samples can be organized into groups according to similarity of composition. As demonstrated in Chapter 4, cluster analysis can be applied and a wide variety of methods are available to explore possible structures and similarities in the analytical data. The result of cluster analysis may be that the samples can be clearly distinguished, by some combination of analyte concentrations, into two groups, and we may wish to use this information to identify and categorize future samples as belonging to one of the two groups. This latter process is *classification*, and the means of deriving the classification rules from previously classified samples is referred to as *discrimination*. It is a pre-requisite for undertaking this supervised pattern recognition that a suitable collection of pre-assigned objects, the *training set*, is available in order to determine the *discriminating rule* or *discriminant function*.[1-3]

The precise nature and form of the classifying function used in a pattern recognition exercise is largely dependent on the analytical data. If the parent population distribution of each group is known to follow the normal curve, then *parametric methods* such as statistical discriminant analysis can be usefully employed. Discriminant analysis is one of the most powerful and commonly used pattern recognition techniques and algorithms are generally available with all commercial statistical software packages. If, on the other hand, the distribution of the data is unknown, or known not to be normal, then *non-parametric methods* come to the fore. One of the most widely used non-parametric algorithms is that of *K-nearest neighbours*.[4] Finally, in recent years, considerable interest has been shown in the use of artificial neural networks for supervised pattern recognition and many examples have been reported in the analytical chemistry literature.[5] In this chapter each of these techniques is examined along with its application to analytical data.

2 Discriminant Functions

The most popular and widely used parametric method for pattern recognition is discriminant analysis. The background to the development and use of this technique will be illustrated using a simple bivariate example.

In monitoring a chemical process, it was found that the quality of the final product can be assessed from spectral data using a simple two-wavelength photometer. Table 1 shows absorbance data recorded at these two wavelengths (400 and 560 nm) from samples of 'good' and 'bad' products, labelled Group A and Group B respectively. On the basis of the data presented, we wish to derive a rule to predict which group future samples can be assigned to, using the two wavelength measures.

Examining the analytical data, the first step is to determine their descriptive statistics, *i.e.* the mean and standard deviation for each variable in Group A and Group B. It is evident from Table 1 that at both wavelengths Group A exhibits higher mean absorbance than samples from Group B. In addition, the standard deviation of data from each variable in both groups is similar. If we consider just one variable, the absorbance at 400 nm, then a first attempt at classification would assign the samples to groups according to this absorbance value. Figure 1 illustrates the predicted effect of such a scheme. The mean

[1] B.K. Lavine, in 'Practical Guide to Chemometrics', ed. S.J. Haswell, Marcel Dekker, New York, USA, 1992.
[2] M. James, 'Classification Algorithms', Collins, London, UK, 1985.
[3] F.J. Manly, 'Multivariate Statistical Methods: A Primer', Chapman and Hall, London, UK, 1991.
[4] A.A. Afifi and V. Clark, 'Computer-Aided Multivariate Analysis', Lifetime Learning, California, USA, 1984.
[5] J. Zupan and J. Gesteiger, 'Neural Networks for Chemists', VCH, Weinheim, Germany, 1993.

Table 1 *Absorbance measurements on two classes of material at* 400 *and* 560 nm

Sample	Good material (Group A): 400 nm	560 nm	*Sample*	Bad material (Group B): 400 nm	560 nm
1	0.40	0.60	12	0.20	0.50
2	0.45	0.45	13	0.20	0.40
3	0.50	0.60	14	0.20	0.30
4	0.50	0.70	15	0.25	0.40
5	0.55	0.65	16	0.25	0.25
6	0.60	0.50	17	0.30	0.30
7	0.60	0.60	18	0.35	0.35
8	0.60	0.70	19	0.40	0.30
9	0.65	0.80	20	0.40	0.20
10	0.70	0.60	21	0.50	0.10
11	0.70	0.80			
Mean	0.568	0.636		0.305	0.310
s	0.098	0.109		0.104	0.112

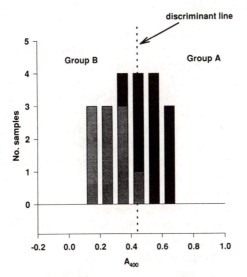

Figure 1 *The distribution of samples from Table 1 according to absorbance measurements at one single wavelength,* 400 nm

values and distribution of the sample absorbances at 400 nm are taken from Table 1, and it is clear that the use of this single variable alone is insufficient to separate the two groups. With the single variable, however, a decision or discriminant function can be proposed.

For equal variances of absorbance data in Groups A and B, the discriminant rule is given by assign sample to Group A if

$$\text{Absorbance}_{400\,\text{nm}} > (\bar{x}_A + \bar{x}_B)/2 \qquad (1)$$

and assign to Group B if

$$\text{Absorbance}_{400\,\text{nm}} < (\bar{x}_A + \bar{x}_B)/2 \qquad (2)$$

i.e. a sample is assigned to the group with the nearest mean value.

Having obtained such a classification rule, it is necessary to test the rule and indicate how good it is. There are several testing methods in common use. Procedures include the use of a set of independent samples or objects not included in the training set, the use of the training set itself, and the *leave-one-out method*. The use of a new, independent set of samples not used in deriving the classification rule may appear the obvious best choice, but it is often not practical. Given a finite size of a data set, such as in Table 1, it would be necessary to split the data into two sets, one for training and one for validation. The problem is deciding which objects should be in which set, and deciding on the size of the sets. Obviously, the more samples used to train and develop the classification rule, the more robust and better the rule is likely to be. Similarly, however, the larger the validation set, the more confidence we can have in the rule's ability to discriminate objects correctly.

The most common method employed to get around this problem is to use all the available data for training the classifier and subsequently test each object as if it were an unknown, unclassified sample. The inherent problem with using the training set as the validation set is that the total classification error, the *error rate*, will be biased low. This is not surprising as the classification rule would have been developed using this same data. New, independent samples may lie outside the boundaries defined by the training set and we do not know how the rule will behave in such cases. This bias decreases as the number of samples analysed increases. For large data sets, say when the number of objects exceeds 10 times the number of variables, the measured *apparent* error can be considered a good approximation of the true error.

If the independent sample set method is considered to be too wasteful of data, which may be expensive to obtain, and the use of the training set for validation is considered insufficiently rigorous, then the leave-one-out method can be employed. By this method all samples but one are used to derive the classification rule, and the sample left out is used to test the rule. The process is repeated with each sample in turn being omitted from the training set and used for validation. The major disadvantage of this method is that there are as many rules derived as there are samples in the data set and this can be computationally demanding. In addition, the error rate obtained refers to the average performance of all the classifiers and not to any particular rule which may subsequently be applied to new, unknown samples.

The results of classification techniques examined in this chapter will be assessed by their apparent error rates using all available data for both training and validation, in line with most commercial software.

Table 2 *Use of the contingency table, or confusion matrix, of classification results* (a). E_{ij} *is the number of objects from group i classified as j,* M_{ia} *is the number of objects actually in group i, and* M_{ic} *is the number classified in group i. The results using the single absorbance at* 400 nm, (b), *and at* 560 nm, (c)

(a)		Actual membership		
		A	*B*	
Predicted	A	E_{AA}	E_{BA}	M_{Ac}
membership	B	E_{AB}	E_{BB}	M_{Bc}
		M_{Aa}	M_{Ba}	

(b)		Actual membership		
		A	*B*	
Predicted	A	10	1	11
membership	B	1	9	10
		11	10	

(c)		Actual membership		
		A	*B*	
Predicted	A	9	1	10
membership	B	2	9	11
		11	10	

The rules expressed by Equations (1) and (2) ensure that the probability of error in misclassifying samples is equal for both groups. In those cases for which the absorbance lies on the discriminant line, samples are assigned randomly to Group A or B. Applying this classification rule to our data results in a total error rate of 9%; two samples are misclassified. To detail how the classifier makes errors, the results can be displayed in the form of a *contingency table*, referred to as a *confusion matrix*, of actual group against classified group, Table 2. A similar result is obtained if the single variable of absorbance at 560 nm is considered alone; three samples are misclassifed.

In Figure 2, the distribution of each variable for each group is plotted along with a bivariate scatter plot of the data and it is clear that the two groups form distinct clusters. However, it is equally evident that it is necessary for both variables to be considered in order to achieve a clear separation. The problem facing us is to determine the best line between the data clusters, the *discriminant function*, and this can be achieved by consideration of probability and *Bayes' theorem*.

Bayes' Theorem

The Bayes' rule simply states that '*a sample or object should be assigned to that group having the highest conditional probability*' and application of this rule to parametric classification schemes provides optimum discriminating capability. An explanation of the term 'conditional probability' is perhaps in order here,

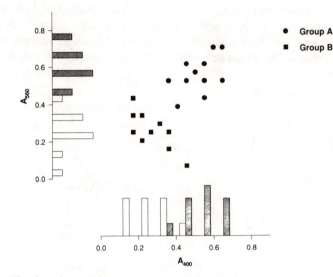

Figure 2 *The data from Table 1 as a scatter plot and, along each axis, the univariate distributions. Two distinct groups are evident from the data*

with reference to a simple example. In spinning a fair coin, the chance of tossing the coin and getting heads is 50%, *i.e.*

$$P_{(heads)} = 0.5 \tag{3}$$

Similarly, the probability of tossing two coins resulting in both showing heads is given by

$$\begin{aligned} P_{(both\,heads)} &= P_{(heads)} \cdot P_{(heads)} \\ P_{(both\,heads)} &= 0.25 \end{aligned} \tag{4}$$

If, however, one coin is already showing heads, then the conditional probability of spinning the other coin and both showing heads is now

$$P_{(both\,heads\,|\,one\,head)} = 0.5 \tag{5}$$

Which is to be read as 'the probability of both coins being heads given that one coin is heads is 0.5', *i.e.* the probability of an event is modified, for better or worse, by prior knowledge.

Of course, if one coin displays tails then,

$$P_{(both\,heads\,|\,one\,tail)} = 0.0 \tag{6}$$

Returning to our analytical problem, of the 21 samples analysed and listed in Table 1, over 50% (11 of the 21) are known to belong to Group A. Thus, in the absence of any analytical data it would seem reasonable to assign any unknown sample to Group A as this has the higher probability of occurrence. With the analytical data presented in Table 1, however, the probability of any sample belonging to one of the groups will be modified according to its absorbance values at 400 and 560 nm. The absorbance values comprise the pattern vectors, denoted by x, where for each sample x_1 is the vector of absorbances at 400 nm and x_2 is the vector of absorbances at 560 nm.

Expressed mathematically, therefore, and applying Bayes' rule, a sample is assigned to Group A, G(A), on the condition that,

$$P_{(G(A)|x)} > P_{(G(B)|x)} \tag{7}$$

Unfortunately, to determine these conditional probability values, *i.e.* confirm that a particular group is characterized by a specific set of variate values, involves the analysis of all potential samples in the parent population. This is obviously unrealistic in practice, and it is necessary to apply Bayes' theorem which provides an indirect means of estimating the conditional probability, $P_{(G(A)|x)}$.

According to Bayes' theorem,

$$P_{(G(A)|x)} = \frac{P_{(x|G(A))} \cdot P_{(G(A))}}{P_{(x|G(A))} \cdot P_{(G(A))} + P_{(x|G(B))} \cdot P_{(G(B))}} \tag{8}$$

$P_{(G(A))}$ and $P_{(G(B))}$ are the *a priori* probabilities, *i.e.* the probabilities of a sample belong to A and B in the absence of having any analytical data.

$P_{(x|G(A))}$ is a conditional probability expressing the chance of a vector pattern x arising from a member of Group A, and this can be estimated by sampling the population of Group A. A similar equation can be arranged for $P_{(G(B)|x)}$ and substitution of Equation (8) into Equation (7) gives

assign sample pattern to Group A if

$$P_{(x|G(A))} \cdot P_{(G(A))} > P_{(x|G(B))} \cdot P_{(G(B))} \tag{9}$$

The denominator term of Equation (8) is common to $P_{(G(A))}$ and $P_{(G(B))}$ and hence cancels from each side of the inequality.

Although $P_{(x|G(A))}$ can be estimated by analysing large numbers of samples, similarly for $P_{(x|G(B))}$, the procedure is still time consuming and requires large numbers of analyses. Fortunately, if the variables contributing to the vector pattern are assumed to possess a multivariate normal distribution, then these conditional probability values can be calculated from

$$P_{(x|G(A))} = \frac{1}{2\pi |Cov|^{1/2}} \exp[-1/2(x - \mu_A)^{\mathrm{T}} \cdot Cov_A^{-1} \cdot (x - \mu_A)] \tag{10}$$

which describes the multidimensional normal distribution for two variables (see Chapter 1). $P_{(x|G(A))}$ can, therefore, be estimated from the vector of Group A mean values, μ_A, and the group covariance matrix, Cov_A.

Substituting Equation (10), and the equivalent for $P_{(x|G(B))}$, in Equation (9), taking logarithms and rearranging leads to the rule

assign sample pattern and object to Group A if,

$$\ln P_{(G(A))} - 0.5\ln(|Cov_A|) - 0.5(x - \mu_A)^T . Cov_A^{-1} . (x - \mu_A) >$$

$$\ln P_{(G(B))} - 0.5\ln(|Cov_B|) - 0.5(x - \mu_B)^T . Cov_B^{-1} . (x - \mu_B) \tag{11}$$

Calculation of the left-hand side of this equation results in a value for each object which is a function of x, the pattern vector, and which is referred to as the *discriminant score*.

The *discriminant function*, $d_A(x)$ is defined by

$$d_A(x) = 0.5\ln(|Cov_A|) + 0.5(x - \mu_A)^T . Cov_A^{-1} . (x - \mu_A) \tag{12}$$

and substituting into Equation (11), the classification rule becomes

assign to Group A if,

$$\ln P_{(G(A))} - d_A(x) > \ln P_{(G(B))} - d_B(x) \tag{13}$$

If the prior probabilities can be assumed to be equal, *i.e.* $P_{(G(A))} = P_{(G(B))}$, then the dividing line between Groups A and B is given by

$$d_A(x) = d_B(x) \tag{14}$$

and Equation (13) becomes

assign object to Group A if,

$$- d_A(x) > - d_B(x)$$

or

$$d_A(x) < d_B(x) \tag{15}$$

The second term in the right-hand side of Equation (12) defining the discriminant function is the quadratic form of a matrix expansion. Its relevance to our discussions here can be seen with reference to Figure 3 which illustrates the division of the sample space for two groups using a simple quadratic function. This Bayes' classifier is able to separate groups with very differently shaped distributions, *i.e.* with differing covariance matrices, and it is commonly referred to as the *quadratic discriminant function*.

The use of Equation (15) can be illustrated by application to the data from Table 1.

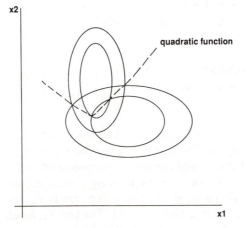

Figure 3 *Contour plots of two groups of bivariate normal data and the quadratic division of the sample space*

From Table 1, the vector of variable means for each group is,

$$\mu_A = \begin{bmatrix} 0.568 \\ 0.636 \end{bmatrix} \qquad \mu_B = \begin{bmatrix} 0.305 \\ 0.310 \end{bmatrix} \tag{16}$$

The variance–covariance matrix for each group can be determined by mean centring the data and pre-multiplying this modified matrix by its transpose, or calculated according to Equation (24) of Chapter 1. Thus,

$$Cov_A = \begin{bmatrix} 0.010 & 0.005 \\ 0.005 & 0.012 \end{bmatrix} \qquad Cov_B = \begin{bmatrix} 0.011 & -0.009 \\ -0.009 & 0.013 \end{bmatrix} \tag{17}$$

and their inverse matrices,

$$Cov_A^{-1} = \begin{bmatrix} 136 & -60 \\ -60 & 109 \end{bmatrix} \qquad Cov_B^{-1} = \begin{bmatrix} 223 & 157 \\ 157 & 190 \end{bmatrix} \tag{18}$$

The determinant of each matrix is

$$|Cov_A| = 8.8 \times 10^{-5} \qquad |Cov_B| = 5.7 \times 10^{-5} \tag{19}$$

The discriminant functions, $d_A(x)$ and $d_B(x)$, for each sample in the training set of Table 1 can now be calculated.

Thus for the first sample,

$$x = \begin{bmatrix} 0.400 \\ 0.600 \end{bmatrix} \quad (x - \mu_A) = \begin{bmatrix} -0.168 \\ -0.036 \end{bmatrix} \quad (x - \mu_B) = \begin{bmatrix} 0.095 \\ 0.290 \end{bmatrix}$$

$$d_A(x) = 0.5 \ln(8.8 \times 10^{-5})$$
$$+ 0.5[-0.168 \quad -0.036] \cdot \begin{bmatrix} 136 & -60 \\ -60 & 109 \end{bmatrix} \cdot \begin{bmatrix} -0.168 \\ -0.036 \end{bmatrix}$$
$$= -3.03$$

$$d_B(x) = 0.5 \ln(5.7 \times 10^{-5})$$
$$+ 0.5[0.095 \quad 0.290] \cdot \begin{bmatrix} 223 & 157 \\ 157 & 190 \end{bmatrix} \cdot \begin{bmatrix} 0.095 \\ 0.290 \end{bmatrix}$$
$$= 8.44 \tag{20}$$

The calculated value for $d_A(x)$ is less than that of $d_B(x)$ so this object is assigned to Group A. The calculation can be repeated for each sample in the training set of Table 1, and the results are provided in Table 3. All 21 samples have been classified correctly as to their parent group. The quadratic discriminating function between the two groups can be derived from Equation (14) by solving the quadratic equations for x. The result is illustrated in Figure 4 and the success of this line in classifying the training set is apparent.

Table 3 *Discriminant scores using the quadratic discriminant function as classifier (a), and the resulting confusion matrix (b)*

Sample	d_A	d_B	Assigned group	Sample	d_A	d_B	Assigned group
1	−3.03	8.44	A	12	2.60	−3.34	B
2	−3.13	2.51	A	13	2.43	−4.38	B
3	−4.43	16.23	A	14	3.36	−3.48	B
4	−3.87	25.76	A	15	0.80	−4.56	B
5	−4.62	25.88	A	16	3.04	−3.69	B
6	−3.32	17.05	A	17	1.03	−4.87	B
7	−4.46	26.25	A	18	−0.68	−4.23	B
8	−4.50	37.35	A	19	0.06	−4.02	B
9	−3.55	57.77	A	20	3.27	−4.38	B
10	−3.12	38.50	A	21	9.17	−2.90	B
11	−3.31	65.74	A				

(b)		Actual group		
		A	B	
Predicted	A	11	0	11
membership	B	0	10	10
		11	10	

Linear Discriminant Function

A further simplification can·be made to the Bayes' classifier if the covariance matrices for both groups are known to be or assumed to be similar. This condition implies that the correlations between variables are independent of the group to which the objects belong. Extreme examples are illustrated in Figure 5. In such cases the groups are linearly separable and a linear discriminant function can be evaluated.

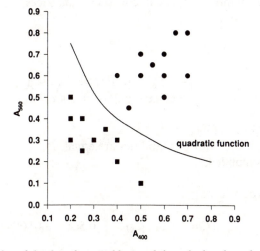

Figure 4 *Scatter plot of the data from Table 1 and the calculated quadratic discriminant function*

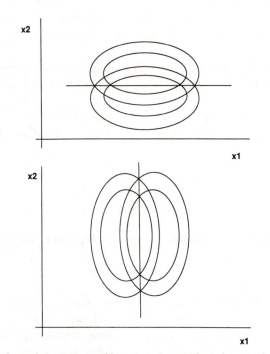

Figure 5 *Contour plots of two groups of bivariate data with each group having identical variance–covariance matrices. Such groups are linearly separable*

With the assumption of equal covariance matrices, the rule defined by Equation (11) becomes

assign to Group A, if,

$$\ln P_{(G(A))} - 0.5(x - \mu_A)^T (Cov^{-1})(x - \mu_A) >$$
$$\ln P_{(G(B))} - 0.5(x - \mu_B)^T (Cov^{-1})(x - \mu_B) \tag{21}$$

where $Cov = Cov_A = Cov_B$.

Once again, if the prior probabilities are equal, $P_{(G(A))} = P_{(G(B))}$, the classification rule is simplified,

assign to Group A, if,

$$- 0.5(x - \mu_A)^T (Cov^{-1})(x - \mu_A) > - 0.5(x - \mu_B)^T (Cov^{-1})(x - \mu_B) \tag{22}$$

which by expanding out the matrix operations simplifies to

assign to Group A if,

$$(\mu_A^T Cov^{-1} x) - 0.5(\mu_A^T Cov^{-1} \mu_A) > (\mu_B^T Cov^{-1} x) - 0.5(\mu_B^T Cov^{-1} \mu_B) \tag{23}$$

Since $\mu_A^T Cov^{-1}$ and $\mu_A^T Cov^{-1} \mu_A$ are constants (they contain no x terms), and similarly $\mu_B^T Cov^{-1}$ and $\mu_B^T Cov^{-1} \mu_B$, then we can define

$$C_{A1} = \mu_A^T Cov^{-1}, \quad C_{B1} = \mu_B^T Cov^{-1}$$
$$C_{A0} = 0.5\mu_A^T Cov^{-1} \mu_A, \quad C_{B0} = 0.5\mu_B^T Cov^{-1} \mu_B \tag{24}$$

and

$$f_A(x) = C_{A1} x - C_{A0}$$
$$f_B(x) = C_{B1} x - C_{B0} \tag{25}$$

The classification rule is now, assign an object to Group A if,

$$f_A(x) > f_B(x) \tag{26}$$

Equations (25) are linear with respect to x and this classification technique is referred to as *linear discriminant analysis*, with the discriminant function obtained by least squares analysis, analogous to multiple regression analysis.

Turning to our spectroscopic data of Table 1, we can evaluate the performance of this linear discriminant analyser.

For the whole set of data and combining all samples from both groups,

$$Cov = \begin{bmatrix} 0.028 & 0.021 \\ 0.021 & 0.040 \end{bmatrix} \quad \text{and} \quad Cov^{-1} = \begin{bmatrix} 60.33 & -32.14 \\ -32.14 & 42.36 \end{bmatrix}$$

and

$$C_{A0} = 0.5 \ \ [0.568 \quad 0.636] \begin{bmatrix} 60.33 & -32.14 \\ -32.14 & 42.36 \end{bmatrix} \begin{bmatrix} 0.568 \\ 0.636 \end{bmatrix} = 0.689$$

$$C_{A1} = [0.568 \quad 0.636] \begin{bmatrix} 60.33 & -32.14 \\ -32.14 & 42.36 \end{bmatrix} = [13.83 \quad 8.68]$$

(27)

$$C_{B0} = 0.5 \ \ [0.305 \quad 0.310] \begin{bmatrix} 60.33 & -32.14 \\ -32.14 & 42.36 \end{bmatrix} \begin{bmatrix} 0.305 \\ 0.310 \end{bmatrix} = 1.803$$

$$C_{B1} = [0.305 \quad 0.310] \begin{bmatrix} 60.33 & -32.14 \\ -32.14 & 42.36 \end{bmatrix} = [8.44 \quad 5.33]$$

Substituting into Equations (25), for the first sample,

$$f_A(x) = [13.83 \quad 8.68] \begin{bmatrix} 0.4 \\ 0.6 \end{bmatrix} - 6.689 = 4.052$$

$$f_B(x) = [8.44 \quad 3.33] \begin{bmatrix} 0.4 \\ 0.6 \end{bmatrix} - 1.803 \ = 3.57$$

(28)

Since the value for $f_A(x)$ exceeds that for $f_B(x)$, from Equation (26) the first sample is assigned to Group A. The remaining samples can be analysed in a similar manner and the results are shown in Table 4. One sample, from Group A, is misclassified. The decision line can be found by solving for x when $f_A(x) = f_B(x)$. This line is shown in Figure 6 and the misclassified sample can be clearly identified.

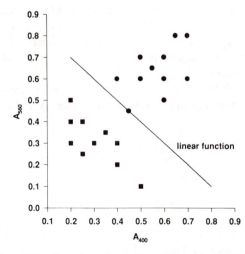

Figure 6 *Scatter plot of the data from Table 1 and the calculated linear discriminant function*

Table 4 *Discriminant scores using the linear discriminant function as classifier (a), and the resulting confusion matrix (b)*

Sample	f_A	f_B	Assigned group	Sample	f_A	f_B	Assigned group
1	4.05	3.57	A	12	0.42	1.55	B
2	3.44	3.49	B	13	− 0.45	1.22	B
3	5.43	4.41	A	14	− 1.32	0.88	B
4	6.30	4.75	A	15	0.24	1.64	B
5	6.54	5.00	A	16	− 1.06	1.14	B
6	5.95	4.92	A	17	0.06	1.73	B
7	6.82	5.26	A	18	1.19	2.31	B
8	7.69	5.59	A	19	1.44	2.57	B
9	9.25	6.34	A	20	0.57	2.24	B
10	8.20	6.10	A	21	1.09	2.75	B
11	9.94	6.77	A				

(b)

		Actual membership A	*Actual membership* B	
Predicted	A	10	0	10
membership	B	1	10	11
		11	10	

As this linear classifier has performed less well than the quadratic classifier, it is worth examining further the underlying assumptions that are made in applying the linear model. The major assumption made is that the two groups of data arise from normal parent populations having similar covariance matrices. Visual examination of Figure 2 indicates that this assumption may not be valid for these absorbance data. The data from samples forming Group A display an apparent positive correlation ($r = 0.54$) between x_1 and x_2, whereas there is negative correlation ($r = − 0.85$) between the absorbance values at the two wavelengths for those samples in Group B. For a more quantitative measure and assessment of the similarity of the two variance–covariance matrices we require some multivariate version of the simple F-test. Such a test may be derived as follows.[6]

For k groups of data characterized by $j = 1 \ldots m$ variables, we may compute k variance–covariance matrices, and for two groups A and B we wish to test the hypothesis

$$H_0: Cov_A = Cov_B$$

against the alternative,

$$H_1: Cov_A \neq Cov_B \qquad (29)$$

If the data arise from a single parent population, then a pooled variance–covariance matrix may be calculated from

[6] J.C. Davis, 'Statistics and Data Analysis in Geology', J. Wiley & Sons, New York, USA, 1973.

$$Cov_p = \sum_{i=1}^{k} \frac{(n_i - 1)\,Cov_i}{\left(\sum_{i=1}^{k} n_i\right) - k} \tag{30}$$

where n_i is the number of objects or samples in group i.

From Equation (30) a statistic, M, can be determined,

$$M = \left[\left(\sum_{i=1}^{k} n_i\right) - k\right] \ln|Cov_p| - \sum_{i=1}^{k} [(n_i - 1)\ln|Cov_i|] \tag{31}$$

which expresses the difference between the logarithm of the determinant of the pooled variance–covariance matrix and the average of the logarithms of the determinants of the group variance–covariance matrices. The more similar the group matrices, the smaller the value of M.

Finally a test statistic based on the χ^2 distribution is generated from

$$\chi^2 = M.C \tag{32}$$

where

$$C = 1 - \frac{2m^2 + 3m - 1}{6(m+1)(k-1)}\left[\sum_{i=1}^{k} \frac{1}{n_i - 1} - \frac{1}{\left(\sum_{i=1}^{k} n_i\right) - k}\right] \tag{33}$$

For small values of k and m, Davis[6] reports that the χ^2 approximation is good, and for our two-group, bivariate sample data the calculation of the χ^2 value is trivial.

$$C = 1 - \frac{(2)(2)^2 + (3)(2) - 1}{6(2+1)(2-1)}\left(\frac{1}{10} + \frac{1}{9} - \frac{1}{21-2}\right) \tag{34}$$

$$= 0.885$$

and

$$M = (21 - 2)\ln|Cov_p| - 10\ln|Cov_A| + 9\ln|Cov_B| \tag{35}$$
$$= 42.1$$

Thus,

$$\chi^2 = 0.885 \times 42.1 = 37.3 \tag{36}$$

with the degrees of freedom given by

$$v = (1/2)(k-1)(m)(m+1) = 3 \tag{37}$$

At a 5% level of significance, the critical value for χ^2 from tables is 7.8. Our value of 37.3 far exceeds this critical value, and the null hypothesis is rejected. We may assume, therefore, that the two groups of samples are unlikely to have similar parent populations and, hence, similar variance–covariance matrices. It is not surprising, therefore, that the linear discriminant analysis model was inferior to the quadratic scheme in classification.

The linear discriminant function is a most commonly used classification technique and it is available with all the most popular statistical software packages. It should be borne in mind, however, that it is only a simplification of the Bayes' classifier and assumes that the variates are obtained from a multivariate normal distribution and that the groups have similar covariance matrices. If these conditions do not hold then the linear discriminant function should be used with care and the results obtained subject to careful analysis.

Linear discriminant analysis is closely related to multiple regression analysis. Whereas in multiple regression, the dependent variable is assumed to be a continuous function of the independent variables, in discriminant analysis the dependent variable, *e.g.* Group A or Group B, is nominal and discrete. Given this similarity, it is not surprising that the selection of appropriate variables to perform a discriminant analysis should follow a similar scheme to that employed in multiple regression (see Chapter 6).

As with multiple regression analysis, the most commonly used selection procedures involve stepwise methods with the *F*-test being applied at each stage to provide a measure of the value of the variable to be added, or removed, in the discriminant function. The procedure is discussed in detail in Chapter 6.

Finally it is worth noting that linear combinations of the original variables may provide better and more effective classification rules than the original variables themselves. Principal components are often employed in pattern recognition and are always worth examining. However, the interpretation of the classification rule in terms of relative importance of variables will generally be more confusing.

3   Nearest Neighbours

The discriminant analysis techniques discussed above rely for their effective use on *a priori* knowledge of the underlying parent distribution function of the variates. In analytical chemistry, the assumption of multivariate normal distribution may not be valid. A wide variety of techniques for pattern recognition not requiring any assumption regarding the distribution of the data have been proposed and employed in analytical spectroscopy. These methods are referred to as non-parametric methods. Most of these schemes are based on attempts to estimate $P_{(x|G_i)}$ and include *histogram techniques*, *kernel estimates* and *expansion methods*. One of the most common techniques is that of *K-nearest neighbours*.

The basic idea underlying nearest-neighbour methods is conceptually very simple, and in practice it is mathematically simple to implement. The general method is based on applying the so-called *K*-nearest neighbour classification rule, usually referred to as *K*-NN. The distance between the pattern vector of

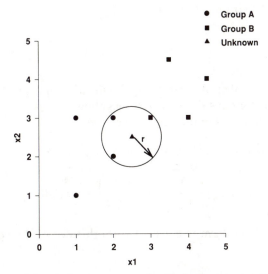

Figure 7 *Radius, r, of the circle about an unclassified object containing three nearest neighbours, two from group A and one from group B. The unknown sample is assigned to group A*

the unclassified sample and every classified sample from the training set is calculated, and the majority of smallest distances, *i.e.* the nearest neighbours, determines to which group the unknown is to be assigned. The most common distance metric used is the Euclidean distance between two pattern vectors.

For objects 1 and 2 characterized by multivariate pattern vectors x_1 and x_2 defined by

$$x_1 = (x_{11}, x_{12}, \ldots, x_{1m})$$
$$x_2 = (x_{21}, x_{22}, \ldots, x_{2m}) \tag{38}$$

where m is the number of variables, the Euclidean distance between objects 1 and 2 is given by

$$d_{12} = \left[\sum_{j=1}^{m} (x_{1i} - x_{2i})^2 \right]^{1/2} \tag{39}$$

Application of Equation (39) to the K-NN rule serves to define a sphere, or circle for bivariate data, about the unclassified sample point in space, of radius r_K which is the distance to the Kth nearest neighbour, containing K nearest neighbours, Figure 7. It is the volume of this sphere which is used as an estimate of $P_{(x|G_i)}$.

For a total training set of N objects comprised of n_i samples known to belong to each group i, the procedure adopted is to determine the Kth nearest neighbour to the unclassified object defined by its pattern vector x, ignoring group membership. From this, the conditional probability of the pattern vector arising from the group i, $P_{(x|G_i)}$, is given by

$$P_{(x|G_i)} = \frac{\Sigma k_i}{n_i} \cdot \frac{1}{V_{K,x}} \qquad (40)$$

where k_i is the number of nearest neighbours in group i and $V_{K,x}$ is the volume of space which contains the K nearest neighbours.

Using Equation (40) in the Bayes' rule gives

assign to group i if

$$P_{(G_i)} \frac{k_i}{n_i} \frac{1}{V_{K,x}} > P_{(G_j)} \frac{k_j}{n_j} \frac{1}{V_{K,x}} \quad \text{for all } j \neq i \qquad (41)$$

Since the volume term is constant to both sides of the equation, the rule simplifies to,

assign to group i if

$$\frac{P_{(G_i)}.k_i}{n_i} > \frac{P_{(G_j)}.k_j}{n_j}, \quad \text{for all } j \neq i \qquad (42)$$

If the number of objects in each training set, n_i, is proportional to the unconditional probability of occurrence of the groups, $P_{(G_i)}$, then Equation (42) simplifies further to

assign to group i if,

$$k_i > k_j \qquad (43)$$

This is a common form of the nearest-neighbour classification rule and assigns a new, unclassified object to that group that contains the majority of its nearest neighbours.

The choice of value for k is somewhat empirical and, for overlapping classes, $k = 3$ or 5 have been proposed to provide good classification. In general, however, $k = 1$ is the most widely used case and is referred to as the 1-NN method or, simply, the nearest-neighbour method.

For our bivariate, spectrophotometric data the inter-sample distance matrix, using the Euclidean metric, is given in Table 5. For each sample the nearest neighbour is highlighted. The confusion matrix summarizes the results, and once again, this time using the 1-NN rule, a single sample from Group A is misclassified.

As well as the Euclidean distance, other metrics have been proposed and employed to measure similarities of pattern vectors between objects. One method used for comparing and classifying spectroscopic data is the *Hamming distance*. For two pattern vectors x_1 and x_2 defined by Equation (38), the

Table 5 *The Euclidean distance matrix for the materials according to the two wavelengths measured* (a), *and the resulting confusion matrix after applying the k-NN classification algorithm*

(a)

1	2	3	4	5	6	7	8	9	10	11	12	13	14	15	16	17	18	19	20	21
0	16	10	14	16	22	20	22	32	30	36	22	28	36	25	38	32	26	30	40	51
16	0	16	26	22	16	21	29	40	29	43	26	26	29	21	28	21	14	16	26	35
10	16	0	10	7	14	10	14	25	20	28	32	36	42	32	43	36	29	32	41	50
14	26	10	0	7	22	14	10	18	22	22	36	42	50	39	51	45	38	41	51	60
16	22	7	7	0	16	7	7	18	16	21	38	43	50	39	50	43	36	38	47	55
22	16	14	22	16	0	10	20	30	14	32	40	41	45	36	43	36	29	28	36	41
20	21	10	14	**7**	**10**	0	10	21	**10**	22	41	45	50	40	50	42	35	36	45	51
22	29	14	10	7	20	10	0	11	14	14	45	50	57	46	57	50	43	45	54	61
32	40	25	18	18	30	21	11	0	21	**5**	54	60	67	57	68	61	54	56	65	72
30	29	20	22	16	14	10	14	21	0	20	51	54	57	49	57	50	43	42	50	54
36	43	28	22	21	32	22	14	**5**	20	0	58	64	71	60	71	64	57	58	67	73
22	26	32	36	38	40	41	45	54	51	58	0	10	20	11	26	22	21	28	36	50
28	26	36	42	43	41	45	50	60	54	64	**10**	0	10	**5**	16	14	16	22	28	42
36	29	42	50	50	45	50	57	67	58	71	20	10	0	11	7	10	16	20	22	36
25	21	32	39	39	36	40	46	57	49	60	11	**5**	11	0	15	11	11	18	25	39
38	28	43	51	50	43	50	57	68	57	71	26	16	**7**	15	0	7	14	16	16	29
32	21	36	45	43	36	42	50	61	50	64	22	14	10	11	**7**	0	7	10	14	28
26	**14**	29	38	36	29	35	43	54	43	57	21	16	16	11	14	**7**	0	7	16	29
30	15	32	41	38	28	36	45	56	42	58	28	22	20	18	16	10	7	0	**10**	22
40	26	41	51	47	36	45	54	65	50	67	36	28	22	25	16	14	16	10	0	**14**
51	35	50	60	55	41	51	61	72	54	73	50	42	36	39	29	28	29	22	14	0

Assigned group
A B A A A A A A A A A B B B B B B B B B B

(b)

		Actual membership		
		A	B	
Predicted	A	10	0	10
membership	B	1	10	11
		11	10	

Hamming distance, H, is simply the absolute difference between each element of one vector and the corresponding component of the other.

$$H = \sum_{i=1}^{m} (|x_{1,i} - x_{2,i}|) \tag{44}$$

When, say, infrared or mass spectra can be reduced to binary strings indicating the presence or absence of peaks or other features, the Hamming distance metric is simple to implement. In such cases it provides a value of differing bits in the binary pattern and is equivalent to performing the exclusive-OR function between the vectors. The Hamming distance is a popular choice in spectral

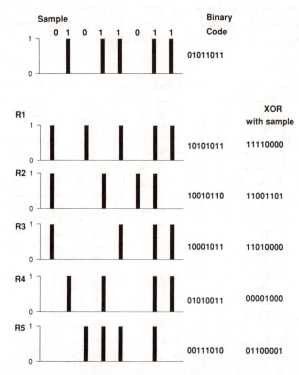

Figure 8 *Binary representation of spectra data (1 – peak, 0 – no peak). The sample has smallest number of XOR bits set with reference spectrum R4, and this, therefore, is the best match*

database 'look-up and compare' algorithms for identifying unknown spectra. Figure 8 provides a simple example of applying the method.

Despite its relative simplicity, the nearest-neighbour classification method often provides excellent results and has been widely used in analytical science. Another advantage of the K-NN technique is that it is a multi-category method. It does not require repeated application to assign some unknown sample to a class as is often the case with binary classifiers. Its major disadvantage is that it is computationally demanding. For each classification decision, the distance between the sample pattern vector and every object in the training set for all groups must be calculated and compared. Where very large training sets are used, however, each distinct class or group can be represented by a few representative patterns to provide an initial first-guess classification before every object in the best classes is examined.

4   The Perceptron

As an approximation to the Bayes' rule, the linear discriminant function provides the basis for the most common of the statistical classification schemes, but there has been much work devoted to the development of simpler linear classification rules. One such method which has featured extensively in spectroscopic pattern recognition studies is the perceptron algorithm.

The perceptron is a simple linear classifier that requires no assumptions to be made regarding the parent distribution of the analytical data. For pattern vectors that are linearly separable, a perceptron will find a hyperplane (in two dimensions this is a line) that completely separates the groups. The algorithm is iterative and starts by placing a line at random in the sample space and examining which side of the line each object in the training set falls. If an object is on the wrong side of the line then the position of the line is changed to attempt to correct the mistake. The next object is examined and the process repeats until a line position is found that correctly partitions the sample space for all objects. The method makes no claims regarding its ability to classify objects not included in the training set, and if the groups in the training set are not linearly separable then the algorithm may not settle to a final stable result.

The perceptron is a learning algorithm and can be considered as a simple model of a biological neuron. It is worth examining here not only as a classifier in its own right, but also as providing the basic features of modern artificial neural networks.

The operation of a perceptron unit is illustrated schematically in Figure 9. The function of the unit is to modify its input signals and produce a binary output, 1 or 0, dependent on the sum of these inputs. Mathematically, the perceptron performs a weighted sum of its inputs, compares this with some threshold value and the output is turned on (output $= 1$) if this value is exceeded, else it remains off (output $= 0$).

For m inputs,

$$\text{total input, } I = \sum_{i=1}^{m} w_i' x_i = w' x' \tag{45}$$

$x' = (x_1 \ldots x_m)$ represents an object's pattern vector, and $w' = (w_1 \ldots w_m)$ is the vector of weights which serve to modify the relative importance of each

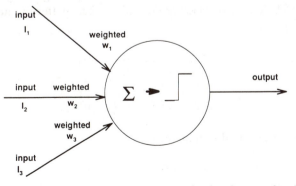

Figure 9 *The simple perceptron unit. Inputs are weighted and summed and the output is '1' or '0' depending on whether or not it exceeds a defined threshold value*

element of x. These weights are varied as the model learns to distinguish between the groups assigned in the training set.

The sum of the inputs, I, is compared with a threshold value, θ, and if $I > \theta$ a value of 1 is output, otherwise 0 is output, Figure 9. This comparison can be achieved by subtracting θ from I and comparing the result with zero, *i.e.* by adding $-\theta$ as an offset to I. The summation and comparison operations can, therefore, be combined by modifying Equation (45),

$$\text{total input, } I = \sum_{i=1}^{m+1} w_i x_i = \boldsymbol{w} \cdot \boldsymbol{x} \tag{46}$$

where now $\boldsymbol{w} = (w_1 \ldots w_{m+1})$ with w_{m+1} being referred to as the unit's bias, and $\boldsymbol{x} = (x_1 \ldots x_{m+1})$ with $x_{m+1} = 1$.

The resulting output, y, is given by

$$y = f_H[\boldsymbol{w} \cdot \boldsymbol{x}] \tag{47}$$

where f_H is the *Heaviside* or step function defined by

$$\begin{aligned} f_H(x) &= 1, \ x > 0 \\ f_H(x) &= 0, \ x \leq 0 \end{aligned} \tag{48}$$

The training of the perceptron as a linear classifier then follows the following steps,

(a) randomly assign the initial elements of the weight vector, $\boldsymbol{w}$,
(b) present an input pattern vector from the training set,
(c) calculate the output value according to Equation (47),
(d) alter the weight vector to discourage incorrect decisions and reduce the classification error,
(e) present the next object's pattern vector and repeat from step (c).

This process is repeated until all objects are correctly classified.

Figure 10(a) illustrates a bivariate data set comprising two groups, each of two objects. These four objects are defined by their pattern vectors, including x_{m+1}, as

$$\begin{aligned} A1, \ \boldsymbol{x} &= [0.2 \quad 0.4 \quad 1.0] \\ A2, \ \boldsymbol{x} &= [0.5 \quad 0.3 \quad 1.0] \\ B1, \ \boldsymbol{x} &= [0.3 \quad 0.7 \quad 1.0] \\ B2, \ \boldsymbol{x} &= [0.8 \quad 0.8 \quad 1.0] \end{aligned} \tag{49}$$

and we take as our initial weight vector

$$\boldsymbol{w} = [1.0 \quad -1.0 \quad 0.5] \tag{50}$$

Thus, our initial partition line, Figure 10(b), is defined by

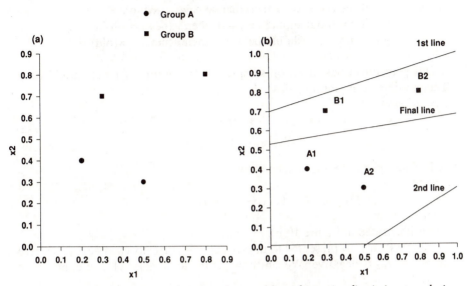

Figure 10 *A simple two-group, bivariate data set* (a), *and iterative discriminant analysis using the simple perceptron* (b)

$$w_1 x_1 + w_2 x_2 + w_3 x_3 = 0$$

i.e.

$$x_1 + 0.5 = x_2 \tag{51}$$

For our first object, A1, the product of x and w is positive and the output is 1, which is a correct result.

$$
\begin{aligned}
I_{A1} &= w.x \\
&= [1 \quad -1 \quad 0.5][0.2 \quad 0.4 \quad 1.0] \\
&= (0.2 \quad -0.4 \quad +0.5) \\
&= 0.3
\end{aligned}
$$

and

$$y_{A1} = f_H(0.3) = 1 \tag{52}$$

For sample A2, the output is also positive and no change in the weight vector is required. For sample B1, however, an output of 1 is incorrect; B1 is not in the same group as A1 and A2, and we need to modify the weight vector. The following weight vector adapting rule is simple to implement,[7]

---

[7] R. Beale and T. Jackson, 'Neural Computing: An Introduction', Adam Hilger, Bristol, UK, 1991.

(a) if the result is correct, then w(new) = w(old),
(b) if $y = 0$ but should be $y = 1$, then w(new) = w(old) + x
(c) if $y = 1$ but should be $y = 0$, then w(new) = w(old) − x.   (53)

Our perceptron has failed on sample B1: the output is 1 but should be 0. Therefore, from Equation (53c),

$$w(\text{new}) = [1 \quad -1 \quad 0.5] - [0.3 \quad 0.7 \quad 1.0]$$
$$= [0.7 \quad -1.7 \quad -0.5] \qquad (54)$$

This new partition line is defined by

$$0.7x_1 - 0.5 = 1.7x_2$$

and is illustrated in Figure 10(b).

Sample B2 is now presented to the system; it is correctly classified with a zero output as I_{B2} is negative. We can now return to sample A1 and continue to

Table 6 *Calculations and results by iteratively applying the perceptron rule to the data illustrated in Figure 10(b)*

Sample	Correct sign		w		$w.x$	Calculated sign	Result
A1	+	1.0	− 1.0	0.5	0.30	+	YES
A2	+				0.70	+	YES
B1	−				0.10	+	NO
B2	−	0.7	− 1.7	− 0.5	− 1.30	−	YES
A1	+				− 1.04	−	NO
A2	+	0.9	− 1.3	0.5	0.56	+	YES
B1	−				− 0.14	−	YES
BS	−				0.18	+	NO
A1	+	0.1	− 2.1	− 0.5	− 1.32	−	NO
A2	+	0.3	− 1.7	0.5	0.14	+	YES
B1	−				− 0.60	−	YES
B2	−				− 0.62	−	YES
A1	+				− 0.12	−	NO
A2	+	0.5	− 1.3	1.5	1.36	+	YES
B1	−				0.74	+	NO
B2	−	0.2	− 2.0	0.5	− 0.94	−	YES
A1	+				− 0.26	−	NO
A2	+	0.4	− 1.6	1.0	0.72	+	YES
B1	−				0.00	?	NO
B2	−	0.1	− 2.3	0	− 1.76	−	YES
A1	+				− 0.9	−	NO
A2	+	0.3	− 1.9	1.0	0.56	+	YES
B1	−				− 0.24	−	YES
B2	−				− 0.28	−	YES
A1	+				0.30	+	YES

repeat the entire process until all samples are classified correctly. The full set of results is summarized in Table 6. The final weight vector is $w = [0.3 \quad -1.9 \quad 1.0]$ with the partition line being

$$0.3x_1 + 1.0 = 1.9x_2 \tag{55}$$

This is illustrated in Figure 10(b) and serves to provide the correct classification of the four objects.

The calculations involved with implementing this perceptron algorithm are simple but tedious to perform manually. Using a simple computer program and analysing the two-wavelength spectral data from Table 1 a satisfactory partition line is obtained eventually and the result is illustrated in Figure 11. The perceptron has achieved a separation of the two groups and every sample has been rightly assigned to its correct parent group.

Several variations of this simple perceptron algorithm can be found in the literature, with most differences relating to the rules used for adapting the weight vector. A detailed account can be found in Beale and Jackson, as well as a proof of the perceptron's ability to produce a satisfactory solution, if such a solution is possible.[7]

The major limitation of the simple perceptron model is that it fails drastically on linearly inseparable pattern recognition problems. For a solution to these cases we must investigate the properties and abilities of multilayer perceptrons and artificial neural networks.

5 Artificial Neural Networks

The simple perceptron model attempts to find a straight line capable of separating pre-classified groups. If such a discriminating function is possible

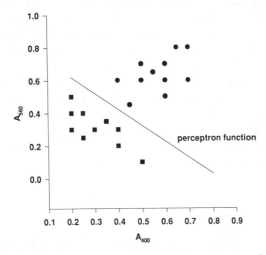

Figure 11 *Partition of the data from Table 1 by a linear function derived from a simple perceptron unit*

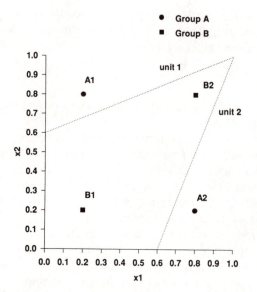

Figure 12 *A simple two-group, bivariate data set that is not linearly separable by a single function. The lines shown are the linear classifiers from the two units in the first layer of the multilayer system shown in Figure 13*

then it will, eventually, be found. Unfortunately, there are many classification problems which are less simple or less tractable.

Consider, for example, the two-group, four-object data set illustrated in Figure 12. Despite the apparent simplicity of this data set, it is immediately apparent that no single straight line can be drawn that will isolate the two classes of sample points. To achieve class separation and develop a satisfactory pattern recognition scheme, it is necessary to modify the simple perceptron.

Correct identification and classification of sets of linearly inseparable items requires two major changes to the simple perceptron model. Firstly, more than one perceptron unit must be used. Secondly, we need to modify the nature of the threshold function. One arrangement which can correctly solve our four-sample problem is illustrated in Figure 13. Each neuron in the first layer receives its inputs from the original data, applies the weight vector, thresholds the weighted sum and outputs the appropriate value of zero or one. These outputs serve as inputs to the second, output layer.

Each perceptron unit in the first layer applies a linear decision function derived from the weight vectors,

$$w_1 = [4 \quad -10 \quad 6]$$
$$w_2 = [-10 \quad 4 \quad 6] \tag{55}$$

which serve to define the lines shown in Figure 12. The weight vector associated with the third, output, perceptron is designed to provide the final classification from the output values of perceptrons 1 and 2,

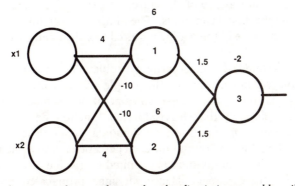

Figure 13 *A two-layer neural network to solve the discriminant problem illustated in Figure 12. The weighting coefficients are shown adjacent to each connection and the threshold or bias for each neuron is given above each unit*

$$w_3 = [1.5 \quad 1.5 \quad -2] \tag{56}$$

We can calculate the output from each perceptron for each sample presented to the input of the system. Thus for object A1,

$$\text{at perceptron 1, } wx = [4 \quad -10 \quad 6][0.2 \quad 0.8 \quad 1]$$
$$= -1.2 \tag{57}$$
$$\therefore \quad y_{p1} = 0$$

$$\text{at perceptron 2, } wx = [-10 \quad 4 \quad 6][0.2 \quad 0.8 \quad 1]$$
$$= 7.2 \tag{58}$$
$$\therefore \quad y_{p2} = 1$$

$$\text{at perceptron 3, } wx = [1.5 \quad 1.5 \quad -2][0 \quad 1 \quad -2]$$
$$= -0.5 \tag{59}$$
$$\therefore \quad y_{p3} = 0$$

Similar calculations can be performed for A2, B1, and B2:

	p1 output	p2 output	p3 output
A1	0	1	0
A2	1	0	0
B1	1	1	1
B2	1	1	1

Perceptron 3 is performing an AND function on the output levels from perceptrons 1 and 2 since its output is 1 only when both inputs are 1.

Although the layout in Figure 13 correctly classifies the data, by applying two linear discriminating functions to the pattern space, it is unable to learn from a training set and must be fully programmed before use, *i.e.* it must be manually set-up before being employed. This situation arises because the second layer is not aware of the status of the original data, only the binary output from the first layer units. The simple on-off output from layer one provides no measure of the scaling required to adjust and correct the weights of its inputs.

The way around the non-learning problem associated with this scheme provides the second change to the simple perceptron model, and involves altering the nature of the comparison operation by modifying the threshold function. In place of the Heaviside step function described previously, a smoother curve such as a linear or sigmoidal function is usually employed, Figure 14. The input and output for each perceptron unit or neuron with such a threshold function will no longer be limited to a value of zero or one, but can range between these extremes. Hence, the signal propagated through the system carries information which can be used to indicate how near an input is to the full threshold value; information which can be used to regulate signal reinforcement by changing the weight vectors. Thus, the multilayer system is now capable of learning.

The basic learning mechanism for networks of multilayer neurons is the *generalized delta rule*, commonly referred to as *back propagation*. This learning rule is more complex than that employed with the simple perceptron unit because of the greater information content associated with the continuous output variable compared with the binary output of the perceptron.

In general, a typical back-propagation network will comprise an input stage, with as many inputs as there are variables, an output layer, and at least one hidden layer, Figure 15. Each layer is fully connected to its succeeding layer. During training for supervised learning, the first pattern vector is presented to the input stage of the network and the output of the network will be unpredictable. This process describes the forward pass of data through the network and, using a sigmoidal transfer function, is defined at each neuron by

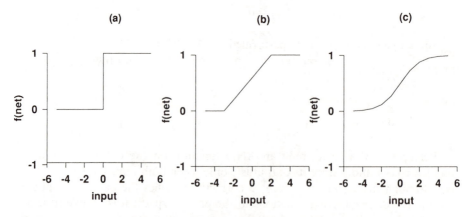

Figure 14 *Some commonly used threshold functions for neural networks: the Heaviside function* (a), *the linear function* (b), *and the sigmoidal function* (c)

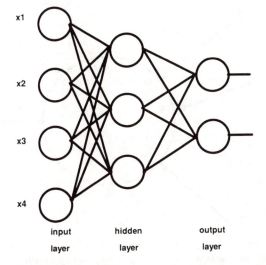

x1

x2

x3

x4

input
layer

hidden
layer

output
layer

Figure 15 *The general scheme for a fully connected two-layer neural network with four inputs*

$$O_j = \frac{1}{1 + e^{-I_j}} \tag{60}$$

where

$$I_j = \sum_i O_i w_{ij} \tag{61}$$

O_j is the output from neuron j and I_j is the summed input to neuron j from other neurons, O_i, modified according to the weight of the connection, w_{ij}, between the i'th and j'th neurons, Figure 16.

The final output from the network for our input pattern is compared with the known, correct result and a measure of the error is computed. In order to reduce this error, the weight vectors between neurons are adjusted by using the generalized delta rule and back-propagating the error from one layer to the previous layer.

The total error, E, is given by the difference between the correct or target output, t, and the actual measured output, O, *i.e.*

$$E = \sum_j (t_j - O_j)^2 \tag{62}$$

and the critical parameter that is passed back through the layers of the network is defined by

$$\delta_j = - \frac{\mathrm{d}E_j}{\mathrm{d}I_j} \tag{63}$$

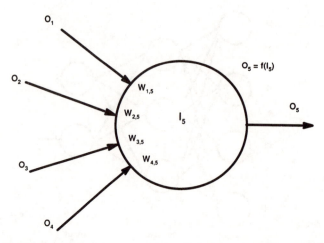

Figure 16 *Considering a single neuron, number 5, in a non-input layer of the network, each of the four inputs, $O_1 \ldots O_4$, is weighted by a coefficient, $w_{15} \ldots w_{45}$. The neuron's output, O_5, is the summed value, I_5, of the inputs modified by the threshold function, $f(I)$*

For output units the observed results can be compared directly with the target result, and

$$\delta_j = f_j'(I_j)(t_j - O_j) \tag{64}$$

where f_j' is the first derivative of the sigmoid, threshold function.

If unit j is not an output unit, then,

$$\delta_j = f_j'(I_j) \sum \delta_k w_k \tag{65}$$

where the subscript k refers to neurons in preceding layers.

Thus the error is calculated first in the output layer and is then passed back through the network to preceding layers for their weight vector to be adapted in order to reduce the error. A discussion of Equations (63) to (65) is provided by Beale and Jackson,[7] and is derived by Zupan.[5]

A suitable neural network can provide the functions of feature extraction and selection and classification. The network can adjust automatically the weights and threshold values of its neurons during a learning exercise with a training set of known and previously categorized data. It is this potential of neural networks to provide a complete solution to pattern recognition problems that has generated the considerable interest in their use. One general problem in applying neural networks relates to the design of the topology of the neural network for any specific problem. For anything other than the most trivial of tasks there may exist many possible solutions and designs which can provide the required classification, and formal rules of design and optimization are rarely employed or acknowledged. In addition, a complex network comprising many hundreds

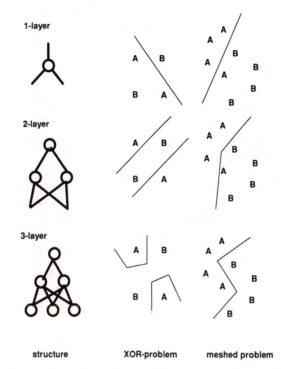

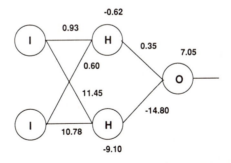

Figure 17   *Neural network configurations and their corresponding decision capabilities illustrated with the XOR problem of Figure 12 and a more complex overlapping 2-group example*
(Reproduced by permission from ref. 7)

Figure 18   *A neural network, comprising an input layer (I), a hidden layer (H), and an output layer (O). This is capable of correctly classifying the analytical data from Table 1. The required weighting coefficients are shown on each connection and the bias values for a sigmoidal threshold function are shown above each neuron*

or thousands of neurons will be difficult, if not impossible, to analyse in terms of its internal behaviour. The performance of a neural network is usually judged by results, often with little attention paid to statistical tests or the stability of the system.

As demonstrated previously, a single-layer perceptron can serve as a linear classifier by fitting a line or plane between the classes of objects, but it fails with non-linear problems. The two-layer device, however, is capable of combining the linear decision planes to solve such problems as that illustrated in Figure 12. Increasing the number of perceptrons or neuron units in the hidden layer increases proportionally the number of linear edges to the pattern shape capable of being classified. If a third layer of neurons is added then even more complex shapes may be identified. Arbitrarily complex shapes can be defined by a three-layer network and such a system is capable of separating any class of patterns. This general principle is illustrated in Figure 17.[6]

For our two-wavelength spectral data, a two-layer network is adequate to achieve the desired separation. A suitable neural network, with the weight vectors, is illustrated in Figure 18.

Calibration and Regression Analysis

1 Introduction

Calibration is one of the most important tasks in quantitative spectrochemical analysis. The subject continues to be extensively examined and discussed in the chemometrics literature as ever more complex chemical systems are studied. The computational procedures discussed in this chapter are concerned with describing quantitative relationships between two or more variables. In particular we are interested in studying how measured *independent* or response variables vary as a function a single so-called *dependent* variable. The class of techniques studied is referred to as *regression analysis*.

The principal aim in undertaking regression analysis is to develop a suitable mathematical model for descriptive or predictive purposes. The model can be used to confirm some idea or theory regarding the relationship between variables or it can be used to predict some general, continuous response function from discrete and possibly relatively few measurements.

The single most common application of regression analysis in analytical laboratories is undoubtedly curve-fitting and the construction of calibration lines from data obtained from instrumental methods of analysis. Such graphs, for example absorbance or emission intensity as a function of sample concentration, are commonly assumed to be linear, although non-linear functions can also be used. The fitting of some 'best' straight line to analytical data provides us with the opportunity to examine the fundamental principles of regression analysis and the criteria for measuring *'goodness of fit'*.

Not all relationships can be adequately described using the simple linear model, however, and more complex functions, such as quadratic and higher-order polynomial equations, may be required to fit the experimental data. Finally, more than one variable may be measured. For example, multiwavelength calibration procedures are finding increasing applications in analytical spectrometry and multivariate regression analysis forms the basis for many chemometric methods reported in the literature.

2 Linear Regression

It frequently occurs in analytical spectrometry that some characteristic, y, of a sample is to be determined as a function of some other quantity, x, and it is necessary to determine the relationship or function between x and y, which may be expressed as $y = f(x)$. An example would be the calibration of an atomic absorption spectrometer for a specific element prior to the determination of the concentration of that element in a series of samples.

A series of n absorbance measurements is made, y_i, one for each of a suitable range of known concentration, x_i. The n pairs of measurements (x_i, y_i) can be plotted as a scatter diagram to provide a visual representation of the relationship between x and y.

In the determination of chromium and nickel in machine oil by atomic absorption spectrometry the calibration data presented in Table 1 were obtained. These experimental data are shown graphically in Figure 1.

At low concentrations of analyte and working at low absorbance values, a linear relationship is to be expected between absorbance and concentration, as predicted by Beer's Law. Visual inspection of Figure 1(a) for the chromium data confirms the correctness of this linear function and, in this case, it is a simple matter to draw by hand a satisfactory straight line through the data and use the plot for subsequent analyses. The equation of the line can be estimated directly from this plot. In this case there is little apparent experimental uncertainty. In many cases, however, the situation is not so clear-cut. Figure 1(b) illustrates the scatter plot of the nickel data. It is not possible here to draw a straight line passing through all points even though a linear relationship

Table 1 *Absorbance data measured from standard solutions of chromium and nickel by AAS (a). Calculation of the best-fit, least-squares line for the nickel data, (b)*

(a)

Chromium concn. (mg kg^{-1}):	0	1	2	3	4	5	(x)
Absorbance:	0.01	0.11	0.21	0.29	0.38	0.52	(y)

Nickel concn. (mg kg^{-1}):	0	1	2	3	4	5	(x)
Absorbance:	0.02	0.12	0.14	0.32	0.38	0.49	(y)

(b)
For Nickel: $x = 2.50$ and $y = 0.245$

							Sum
$(x_i - x)$:	− 2.50	− 1.50	− 0.50	0.50	1.50	2.50	
$(y_i - y)$:	− 0.225	− 0.125	− 0.105	0.075	0.135	0.245	
$(x_i - x)(y_i - y)$:	0.562	0.187	0.052	0.037	0.202	0.621	1.655
$(x_i - x)^2$:	6.25	2.25	0.25	0.25	2.25	0.25	17.50

$$b = 1.655/17.50 = 0.095$$
$$a = 0.0075$$

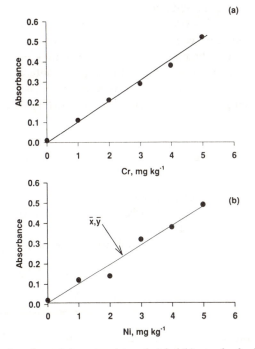

Figure 1 *Calibration plots of chromium (a) and nickel (b) standard solutions, from data in Table 1. For chromium, a good fit can be drawn by eye. For nickel, however, a regression model should be derived, Table 1(b)*

between absorbance and concentration is still considered valid. The deviations in the absorbance data from expected, ideal values can be assumed to be due to experimental errors and uncertainties in the individual measurements and not due to some underlying error in the theoretical relationship. If multiple measurements of absorbance were made for each standard concentration, then a normal distribution for the absorbance values would be expected. These values would be centred on some mean absorbance value $\bar{y}_i$. The task for an analyst is to determine the 'best' straight line regressed through the means of the experimental data.

The data set comprises pairs of measurements of an independent variable x (concentration) and a dependent variable y (absorbance) and it is required to fit the data using a linear model with the well known form,

$$\hat{y}_i = a + bx_i \tag{1}$$

where a and b are constant coefficients characteristic of the regression line, representing the intercept of the line with the y-axis and the gradient of the line respectively. The values of $\hat{y}$ represent the estimated, model values of absorbance derived from this linear model. The generally accepted requirement for deriving the best straight line between x and $\hat{y}$ is that the discrepancy between the measured data and the fitted line is minimized. The most popular technique employed to minimize this error between model and recorded data is the *least-squares* method. For each measured value, the deviation between the derived model value and the measured data is given by $\hat{y}_i - y_i$.

The total error between the model and observed data is the sum of these individual errors. Each error value is squared to make all values positive and prevent negative and positive errors from cancelling. Thus the total error, ϵ, is given by,

$$\text{error,} \quad \epsilon = \sum_{i=1}^{n} (\hat{y}_i - y_i)^2 \tag{2}$$

The total error is the sum of the squared deviations. For some model defined by coefficients a and b, this error will be a minimum and this minimum point can be determined using partial differential calculus.

From Equations (1) and (2) we can substitute our model equation into the definition of error,

$$\epsilon = \sum_{i=1}^{n} (a + bx_i - y_i)^2 \tag{3}$$

The values of the coefficients a and b are our statistically independent unknowns to be determined. By differentiating with respect to a and b respectively, then at the minimum,

$$\frac{\delta\epsilon}{\delta a} = \sum_{i=1}^{n} 2(\hat{a} + \hat{b}x_i - y_i)^2 = 0$$

$$\frac{\delta\epsilon}{\delta b} = \sum_{i=1}^{n} 2x_i(\hat{a} + \hat{b}x_i - y_i)^2 = 0 \tag{4}$$

where $\hat{a}$ and $\hat{b}$ are least squares estimates of the intercept, a, and slope, b.

Expanding and rearranging Equations (4) provides the two simultaneous equations,

$$n\hat{a} + \hat{b}\Sigma x_i = \Sigma y_i$$
$$\hat{a}\Sigma x_i + \hat{b}\Sigma x_i^2 = \Sigma(y_i x_i) \tag{5}$$

from which the following expressions can be derived,

$$\hat{a} = \bar{y} - \hat{b}\bar{x} \tag{6}$$

and

$$\hat{b} = \frac{\Sigma(x_i - \bar{x})(y_i - \bar{y})}{\Sigma(x_i - \bar{x})^2} \tag{7}$$

where $\bar{x}$ and $\bar{y}$ represent the mean values of x and y.[1]

For the experimental data for Ni, calculation of $\hat{a}$ and $\hat{b}$ is trivial ($a = 0.0075$ and $b = 0.095$) and the fitted line passes through the central point given by $\bar{x}$, $\bar{y}$, Figure 1(b).

Once values for $\hat{a}$ and $\hat{b}$ are derived, it is possible to deduce the concentration of subsequently analysed samples by recording their absorbances and substituting the values in Equation (1). It should be noted, however, that because the model is derived for concentration data in the range defined by x_i it is important that subsequent predictions are also based on measurements in this range. The model should be used for interpolation only and not extrapolation.

Errors and Goodness of Fit

It is often the case in chemical analysis that the independent variable, standard solution concentrations in the above example, is said to be *fixed*. The values of concentration for the calibration solutions can be expected to have been chosen by the analyst and the values to be accurately known. The errors associated with x, therefore, are negligible compared with the uncertainty in y due to fluctuations and noise in the instrumental measurement.

To use Equations (6) and (7) in order to determine the characteristics of the fitted line, and employ this information for prediction, it is necessary to estimate the uncertainty in the calculated values for the slope, $\hat{b}$, and intercept, $\hat{a}$. Each of the absorbance values, y_i, has been used in the determination of $\hat{a}$ and $\hat{b}$ and each has contributed its own uncertainty or error to the final result.

Estimates of error in the fitted line and estimates of confidence intervals may be made if three assumptions are valid,

(a) the absorbance values are from parent populations normally distributed about the mean absorbance value,
(b) the variance associated with absorbance is independent of absorbance, *i.e.* the data are *homoscedastic*, and,
(c) the sample absorbance means lie on a straight line.

These conditions are illustrated in Figure 2 which illustrates a theoretical regression line of such data on an independent variable.

The deviation or *residual* for each of the absorbance values in the nickel data is given by $y_i - \hat{y}_i$, *i.e.* the observed values minus the calculated or predicted values according to the linear model. The sum of the squares of these deviations, Table 2, is the *residual sum of squares*, and is denoted as SS_D. The least squares estimate of the line can be shown to provide the best possible fit and no other line can be fitted that will produce a smaller sum of squares.

$$SS_D = \epsilon = \Sigma(y_i - \hat{y}_i)^2 = 0.00599 \tag{8}$$

The variance associated with these deviations will be given by this sum of squares divided by the number of degrees of freedom,

[1] C. Chatfield, 'Statistics for Technology', Chapman and Hall, London, UK, 1975.

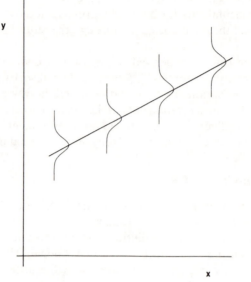

Figure 2 *A regression line through mean values of homoscedastic data*

$$\sigma_D{}^2 = SS_D/(n-2) = \Sigma(y_i - \hat{y}_i)^2/(n-2) \tag{9}$$

The denominator, $n-2$, is the residual degrees of freedom derived from the sample size, n, minus the number of parameters estimated for the line, $\hat{a}$ and $\hat{b}$.

The standard deviations or errors of the estimated intercept and slope values, denoted by σ_a and σ_b respectively, are defined by[2]

$$\sigma_a = \sigma_D \left[\frac{1}{n} \frac{\Sigma x_i^2}{\Sigma(x_i - \bar{x})^2} \right]^{0.5}$$

$$\sigma_b = \sigma_D / [\Sigma(x_i - \bar{x})^2]^{0.5} \tag{10}$$

from which the confidence intervals, CI, can be obtained,

$$CI(a) = \hat{a} \pm t\sigma_a$$

and

$$CI(b) = \hat{b} \pm t\sigma_b \tag{11}$$

where t is the value of the two-tailed t-distribution with $(n-2)$ degrees of freedom. Table 2 gives the results for $CI(a)$ and $CI(b)$ using 95% confidence intervals for the nickel absorbance data.

[2] J.N. Miller, *Analyst*, 1991, **116**, 3.

Table 2 *Errors and goodness of fit calculations associated with the linear regression model for nickel AAS data from Table 1*

Nickel concn. (mg kg^{-1}):	0	1	2	3	4	5
Absorbance (measured):	0.02	0.12	0.14	0.32	0.38	0.49
Absorbance (estimated):	0.007	0.102	0.197	0.292	0.387	0.482

$SS_D = 0.00599$

$s_D = 0.0015$ $\qquad s_a = 0.028$ $\qquad s_b = 0.0093$

$CI(a) = 0.0075 + / - 0.078$ $\qquad CI(b) = 0.095 + / - 0.026$

$SS_T = 0.177$ $\qquad SS_R = 0.171$ $\qquad r^2 = 0.966$

How well the estimated straight line fits the experimental data can be assessed by determining the *coefficient of determination* and the correlation coefficient.

The total variation associated with the y values, SS_T, is given by the sum of the squared deviations of the observed y values from the mean y value,

$$SS_T = \Sigma(y_i - \bar{y})^2 \tag{12}$$

This total variation comprises two components, that due to the residual or deviation sum of squares, SS_D, and that from the sum of squares due to regression, SS_R:

$$SS_T = SS_D + SS_R \tag{13}$$

SS_D is a measure of the failure of the regressed line to fit the data points, and SS_R provides a measure of the variation in the regression line about the mean values.

The ratio of SS_R to SS_T indicates how well the model straight line fits the experimental data. It is referred to as the coefficient of determination and its value varies between zero and one. From Equation (13), if $SS_D = 0$ (the fitted line passes through each datum point) the total variation in y is explained by the regression line and $SS_T = SS_R$ and the ratio is one. On the other hand, if the regressed line fails completely to fit the data, SS_R is zero, the total error is dominated by the residuals, *i.e.* $SS_T \sim SS_D$, then the ratio is zero and no linear relationship is present in the data.

The coefficient of determination is denoted by r^2,

$$r^2 = SS_R/SS_T \tag{14}$$

and r^2 is the square of the correlation coefficient, r, introduced in Chapter 1.

From our data of measured absorbance *vs.* nickel concentration, $r^2 = 0.966$, indicating a good fit between the linear model and the experimental model. As discussed in Chapter 1, however, care must be taken in relying too much on high values of r^2 or r as indicators of linear trends. The data should be plotted and examined visually.

In quantitative spectroscopic analysis an important parameter is the estimate of the confidence interval of a concentration value of an unknown sample, x_u, derived from a measured instrument response. This is discussed in detail by Miller[2] and can be obtained from the standard deviation associated with x_u,

$$\sigma_{x(u)} = \frac{\sigma_D}{b}\left[\frac{1}{m} + \frac{1}{n} + \frac{(y_u - \bar{y})^2}{b^2\Sigma(x_i - \bar{x})^2}\right]^{0.5} \tag{15}$$

where y_u is the mean absorbance of the unknown sample from m measurements. Thus, from a sample having a mean measured absorbance of 0.25 (from five observations),

$$\sigma_{x(u)} = 0.248 \tag{16}$$

and the 95% confidence limits of x_u are

$$CI(x_u) = 2.55 \pm 0.69 \tag{17}$$

Regression through the Origin

Before leaving linear regression, a special case often encountered in laboratory calibrations should be considered. A calibration is often performed using not only standard samples containing known amounts of the analyte but also a blank sample containing no analyte. The measured response for this blank sample may be subtracted from the response values for each standard sample and the fitted line assumed, and forced, to pass through the origin of the graph.

Under such conditions the estimated regression line, Equation (1), reduces to

$$y_i = \hat{b}x_i$$

and

$$\epsilon = \Sigma(\hat{b}x_i - y_i)^2 \tag{18}$$

The resulting equation for b, following partial differentiation, is

$$\hat{b} = \Sigma(x_iy_i)/\Sigma x_i^2 \tag{19}$$

The option to use this model is often available in statistical computer packages, and for manual calculations the arithmetic is reduced compared with the full linear regression discussed above. A caveat should be made, however, since forcing the line through the origin assumes that the measured blank value is free from experimental error and that it represents accurately the true, mean blank value.

For the nickel data from Table 1, using Equation (19), $b = 0.094$, and the sum of squares of the deviations, SS_D, is 0.00614. This value is greater than the computed value of SS_D for the model using data not corrected for the blank, indicating the poorer performance of the model of Equation (18).

3 Polynomial Regression

Although the linear model is the model most commonly encountered in analytical science, not all relationships between a pair of variables can be adequately described by linear regression. A calibration curve does not have to approximate a straight line to be of practical value. The use of higher-order equations to model the association between dependent and independent variables may be more appropriate. The most popular function to model non-linear data and include curvature in the graph is to fit a power-series polynomial of the form

$$y = a + bx + cx^2 + dx^3 + \ldots \tag{20}$$

where, as before, y is the dependent variable, x is the independent variable to be regressed on y, and a, b, c, d, *etc.* are the coefficients associated with each power term of x.

The method of least squares was employed in the previous section to fit the best straight line to analytical data and a similar procedure can be adopted to estimate the best polynomial line. To illustrate the technique, the least squares fit for a quadratic curve will be developed. This can be readily extended to higher power functions.[1,3]

The quadratic function is given by

$$y = a + bx + cx^2 \tag{21}$$

and the following simultaneous equations can be derived:

$$\begin{aligned}
a\Sigma 1 + b\Sigma x + c\Sigma x^2 &= \Sigma y \\
a\Sigma x + b\Sigma x^2 + c\Sigma x^3 &= \Sigma yx \\
a\Sigma x^2 + b\Sigma x^3 + c\Sigma x^4 &= \Sigma yx^2
\end{aligned} \tag{22}$$

and in matrix notation,

$$\begin{bmatrix} n & \Sigma x & \Sigma x^2 \\ \Sigma x & \Sigma x^2 & \Sigma x^3 \\ \Sigma x^2 & \Sigma x^3 & \Sigma x^4 \end{bmatrix} \begin{bmatrix} a \\ b \\ c \end{bmatrix} = \begin{bmatrix} \Sigma y \\ \Sigma yx \\ \Sigma yx^2 \end{bmatrix} \tag{23}$$

which can be solved for coefficients a, b, and c.

The extension of the technique to higher order polynomials, *e.g.* cubic, quartic, *etc.*, is straightforward. Consider the general m'th degree polynomial

[3] A.F. Carley and P.H. Morgan, 'Computational Methods in the Chemical Sciences', Ellis Horwood, Chichester, UK, 1989.

$$y = a + bx + cx^2 + \ldots + zx^m \tag{24}$$

This expands to $(m + 1)$ simultaneous equations from which $(m + 1)$ co-efficients are to be determined. The terms on the right-hand side of the matrix equation will range from Σy_i to $\Sigma(x_i^m . y_i)$ and on the left-hand side from $\Sigma 1$ to Σx_i^{2m}.

A serious problem encountered with the application of polynomial curve-fitting is the fundamental decision as to which degree of polynomial is best. Visual inspection of the experimental data may indicate that a straight line is not appropriate. It may not be immediately apparent, unless theory dictates otherwise, whether say, a quadratic or cubic equation should be employed to model the data. As the number of terms in the polynomial is increased, the correlation between the experimental data and the fitted curve will also increase. In the limit, when the number of terms is one less than the number of the data points the correlation will be unity, *i.e.* the curve will pass through every point. Such a polynomial, however, may have no physical significance. In practice, statistical tests, based on the use of the *F*-ratio, can be employed to examine the significance of terms added to a polynomial and to indicate whether observed increases in the correlation coefficient are statistically significant.

Table 3 shows results of recorded fluorescence emission intensity as a function of concentration of quinine sulphate in acidic solutions. These data are plotted in Figure 3 with regression lines calculated from least squares estimated lines for a linear model, a quadratic model and a cubic model. The correlation for each fitted model with the experimental data is also given. It is obvious by visual inspection that the straight line represents a poor estimate of the association between the data despite the apparently high value of the correlation coefficient. The observed lack of fit may be due to random errors in the measured dependent variable or due to the incorrect use of a linear model. The latter is the more likely cause of error in the present case. This is confirmed by examining the differences between the model values and the actual results, Figure 4. With the linear model, the residuals exhibit a distinct pattern as a function of concentration. They are not randomly distributed as would be the case if a more appropriate model was employed, *e.g.* the quadratic function.

The linear model predicts the relationship between fluorescence intensity, I, and analyte concentration, x, to be of the form,

$$I_i = a + bx_i + \epsilon_i \tag{25}$$

Table 3 *Measured fluorescence emission intensity as a function of quinine concentration*

Quinine concn. (mg kg^{-1}):	0	5	10	15	20	25
Fluorescence intensity: (arb. units)	10	180	300	390	460	520

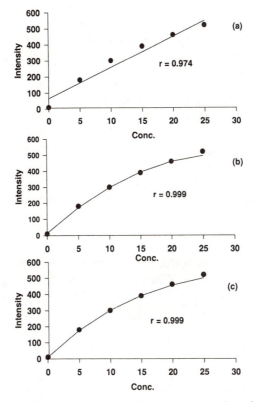

Figure 3 *The linear (a), quadratic (b), and cubic (c) regression lines for the fluorescence data from Table 3*

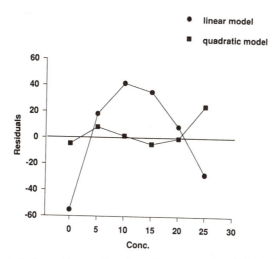

Figure 4 *Residuals ($y_i - \hat{y}_i$) as a function of concentration (x) for best fit linear and quadratic models*

where ϵ is a random error, assumed to be normally distributed, with a variance, σ^2, independent of the value of I. If these assumptions are valid and Equation (25) is a true model of the experimental data then the variance of ϵ will be equal to the variance about the regression line. If the model is incorrect, then the variance around the regression will exceed the variance of ϵ. These variances can be estimated using ANOVA and the F-ratio calculated to compare the variances and test the significance of the model.

The form of the ANOVA table for multiple regression is shown in Table 4. The completed table for the linear model fitted to the fluorescence data is given in Table 5. This analysis of variance serves to test whether a regression line is helpful in predicting the values of intensity from concentration data. For the linear model we wish to test whether the line of slope b adds a significant contribution to the zero-order model. The null hypothesis being tested is,

$$H_0: b = 0 \tag{26}$$

i.e. the mean concentration value is as accurate in predicting emission intensity as the linear regression line. When the fitted line differs significantly from a horizontal ($b = 0$) line, then the term $\Sigma(\bar{I} - \hat{I}_i)^2$ will be large relative to the

Table 4 *ANOVA table for multiple regression*

Source of variation	Sum of Squares (SS)	Degrees of freedom (df)	Mean Squares (MS)	F-ratio
Regression	$(\bar{y} - \hat{y}_i)^2$	p	SS_{reg}/p	MS_{reg}/MS_{res}
Residuals (error)	$(y_i - \hat{y}_i)^2$	$n - p - 1$	$SS_{res}/(n - p - 1)$	
Total	$(y_i - \bar{y})^2$	$n - 1$	$SS_{tot}/(n - 1)$	

Table 5 *ANOVA table for the linear regression model applied to the fluorescence data, emission intensity as a function of concentration*

Source of variation	Sum of Squares	df	Mean Squares	F-ratio
Regression	163 206	1	163 206	74.8
Residuals	8728	4	2182	
Total	171 934	5	34 388	

$$I = 65.24 + 19.31x$$
$$r^2 = 0.949$$

Table 6 *ANOVA table for the quadratic regression model of fluorescence intensity as a function of concentration*

Source of variation	Sum of Squares	df	Mean Squares	F-ratio
Regression	171 807	2	85 903	2038
Residuals	126	3	42	
Total	171 934	5	34 387	

$$I = 14.64 + 34.49x - 0.61x^2$$
$$r^2 = 0.999$$

residuals from the line, $\Sigma(I_i - \hat{I}_i)^2$. As expected, this in fact is the case for the linear model, $F_{1,4} = 74.8$, compared with $F_{1,4} = 7.71$ from tables for a 5% level of significance. So the null hypothesis is rejected, the linear regression model is significant, and the degree to which the regression equation fits the data can be evaluated from the coefficient of determination, r^2, given by Equation (14).

A similar ANOVA table can be completed for the quadratic model, Table 6. Does the addition of a quadratic term contribute significantly to the first-order, linear model? The equation tested is now

$$I_i = a + bx_i + cx_i^2 + \epsilon_i \tag{27}$$

and the null hypothesis is

$$H_0: b = c = 0 \tag{28}$$

Once again the high value of the F-ratio indicates the model is significant as a predictor. This analysis can now be taken a step further since the sum of the squares associated with the regression line can be attributed to two components, the linear function and the quadratic function. This analysis is accomplished by the decomposition of the sum of squares, Table 7. The total sum of squares values for the regression can be obtained from Table 6 and that due to

Table 7 *Sum of Squares decomposition for the quadratic model*

Source of variation	Sum of Squares	df	Mean Squares	F-ratio
x	163 206	1	163 206	3873
x^2	8601	1	8601	204
Total	171 807	2	85 903	

Table 8 *ANOVA table for the cubic regression model of fluorescence intensity as a function of concentration*

Source of variation	Sum of Squares	df	Mean Squares	F-ratio
Regression	171 901	3	57 300	3522
Residuals	32	2	16	
Total	171 933	5	34 387	

$$I = 11.03 + 37.8x - 0.97x^2 + 0.0096x^3$$
$$r^2 = 0.999$$

Table 9 *Sum of Squares decomposition for the cubic model*

Source of variation	Sum of Squares	df	Mean Squares	F-ratio
x	163 206	1	163 206	10 031
x^2	8601	1	8061	529
x^3	94	1	94	5.8
Total	171 901	3	57 300	

the linear component, x, from Table 5. The difference is attributed to the quadratic term. The large F-value indicates the high significance of each term.

The exercise can be repeated for the fitted cubic model, and the ANOVA table and sums of squares decomposition are shown in Tables 8 and 9 respectively. In this case, the F-statistic for the cubic term ($F = 5.8$) is not significant at the 5% level. The cubic term is not required and we can conclude that the quadratic model is sufficient to describe the analytical data accurately, a result which agrees with visual inspection of the line, Figure 3(b).

In summary the three models tested are

$$I = 65.24 + 19.31x$$
$$I = 14.64 + 34.49x - 0.61x^2 \tag{29}$$
$$I = 11.03 + 38.80x - 0.97x^2 + 0.0096x^3$$

The relative effectiveness and importance of the variables can be estimated from the relative magnitudes of the regression coefficients. This cannot be done directly on these coefficients, however, as their magnitudes are dependent on the magnitudes of the variables themselves. In Equation (29), for example, the coefficient for the cubic term is small compared with those for the linear and quadratic terms, but the cubic term itself may be very large. Instead, the *standardized regression coefficients, B_i*, are employed. These are determined by

$$B_k = b_k \sigma_k / \sigma_y \qquad (30)$$

where σ_k is the standard deviation of the variable x_k and σ_y is the standard deviation of the dependent variable, y.

For the cubic model,

$$
\begin{aligned}
B_1 &= b\sigma_x/\sigma_y = 37.8\,(9.35/185.4) = 1.91 \\
B_2 &= c\sigma_x 2/\sigma_y = -0.97\,(243.6/185.4) = 1.27 \\
B_3 &= d\sigma_x 3/\sigma_y = 0.0096\,(6143.5/185.4) = 0.32
\end{aligned}
\qquad (31)
$$

As expected, the relative significance of the standard regression coefficient B_3 is considerably less than those of the standardized linear and quadratic coefficients, B_1 and B_2.

Orthogonal Polynomials

In the previous section, the fluorescence emission data were modelled using linear, quadratic, and cubic equations and the quadratic form was determined as providing the most appropriate model. Despite this, on moving to the higher, cubic, polynomial the coefficient of the cubic term is not zero and the values for the regression coefficients are considerably different from those obtained for the quadratic equation. In general, the least squares polynomial fitting procedure will yield values for the coefficients which are dependent on the degree of the polynomial model. This is one of the reasons why the use of polynomial curve fitting often contributes little to understanding the causal relationship between independent and dependent variables, despite the technique providing a useful curve fitting procedure.

With the general polynomial equation discussed above, the value of the first coefficient, a, represents the intercept of the line with the y-axis. The b coefficient is the slope of the line at this point, and subsequent coefficients are the values of higher orders of curvature. A more physically significant model might be achieved by modelling the experimental data with a special polynomial equation; a model in which the coefficients are not dependent on the specific order of equation used. One such series of equations having this property of independence of coefficients is that referred to as *orthogonal polynomials*.

Bevington[4] presents the general orthogonal polynomial between variables y and x in the form

$$y = a + b(x - \beta) + c(x - \gamma_1)(x - \gamma_2) + d(x - \delta_1)(x - \delta_2)(x - \delta_3) + \dots \qquad (32)$$

As usual, the least squares procedure is employed to determine the values of the regression coefficients a, b, c, d, *etc.*, giving the minimum deviation between the observed data and the model. Also, we impose the criterion that subsequent

---

4 P.R. Bevington, 'Data Reduction and Error Analysis in the Physical Sciences', McGraw-Hill, New York, USA, 1969.

addition of higher-order terms to the polynomial will not change the value of the coefficients of lower-order terms. This extra constraint is used to evaluate the parameters β, γ_1, γ_2, δ_1, *etc*. The coefficient a represents the average y value, b the average slope, c the average curvature, *etc*.

In general, the computation of orthogonal polynomials is laborious but the arithmetic can be greatly simplified if the values of the independent variable, x, are equally spaced and the dependent variable is homoscedastic.[4] In this case,

$$\beta = \bar{x}$$

$$\gamma = \beta \pm \Delta \left[\frac{n^2 - 1}{12} \right]^{0.5}$$
(33)

$$\delta = \beta, \beta \pm \Delta \left[\frac{3n^2 - 7}{20} \right]^{0.5}$$

where

$$\Delta = x_{i+1} - x_i.$$
(34)

and

$$a = \bar{y}$$

$$b = \frac{\Sigma y_i(x_i - \beta)}{\Sigma(x_i - \beta)^2}$$

$$c = \frac{\Sigma[\, y_i(x_i - \gamma_1)(x_i - \gamma_2)]}{\Sigma[(x_i - \gamma_1)(x_i - \gamma_2)]^2}$$
(35)

$$d = \frac{\Sigma[\, y_i(x_i - \delta_1)(x_i - \delta_2)(x_i - \delta_3)]}{\Sigma[(x_i - \delta_1)(x_i - \delta_2)(x_i - \delta_3)]^2}$$

Orthogonal polynomials are particularly useful when the order of the equation is not known beforehand. The problem of finding the lowest-order polynomial to represent the data adequately can be achieved by first fitting a straight line, then a quadratic curve, then a cubic, and so on. At each stage it is only necessary to determine one additional parameter and apply the F-test to estimate the significance of each additional term.

For the fluorescence emission data,

$$\beta = 12.5$$
$$\gamma_1 = 21.04, \gamma_2 = 3.96$$
$$\delta_1 = 12.5, \delta_2 = 23.74, \delta_3 = 1.26$$
(36)

and, from Equations (34),

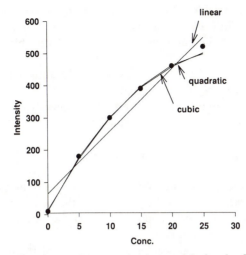

Figure 5 *Orthogonal linear, quadratic, and cubic models for the fluorescence intensity data from Table 3*

$$a = 306.7$$
$$b = 19.31$$
$$c = -0.608$$ (37)
$$d = 0.0089$$

Thus the orthogonal linear equation is given by

$$I_i = 306.7 + 19.31(x_i - 12.5)$$

the quadratic by

$$I_i = 306.7 + 19.31(x_i - 12.5) - 0.608(x_i - 21.04)(x_i - 3.96)$$

and the cubic model by

$$I_i = 306.7 + 19.31(x_i - 12.5) - 0.608(x_i - 21.04)(x_i - 3.96)$$
$$+ 0.0089(x_i - 12.5)(x_i - 23.74)(x_i - 1.26)$$ (38)

These equations are illustrated graphically in Figure 5. As before, an ANOVA table can be constructed for each model and the significance of each term estimated by sums of squares decomposition and comparison of standard regression coefficients.

4 Multivariate Regression

To this point, the discussion of regression analysis and its applications has been limited to modelling the association between a dependent variable and a single independent variable. Chemometrics is more often concerned with multi-variate measures. Thus it is necessary to extend our account of regression to include cases in which several or many independent variables contribute to the measured response. It is important to realize at the outset that the term independent variables as used here does not imply statistical independence, as the x variables may be highly correlated.

In the simplest example, the dependent response variable, y, may be a function of two such independent variables, x_1 and x_2.

$$y = a + b_1 x_1 + b_2 x_2 \qquad (39)$$

Again a is the intercept on the ordinate y-axis, and b_1 and b_2 are the *partial regression coefficients*. These coefficients denote the rate of change of the mean of y as a function of x_1, with x_2 constant, and the rate of change of y as a function of x_2 with x_1 constant.

Multivariate regression analysis plays an important role in modern process control analysis, particularly for quantitative UV–visible absorption spectrometry and near-IR reflectance analysis. It is common practice with these techniques to monitor absorbance, or reflectance, at several wavelengths and relate these individual measures to the concentration of some analyte. The results from a simple two-wavelength experiment serve to illustrate the details of multivariate regression and its application to multivariate calibration procedures.

Figure 6 presents a UV spectrum of the amino acid tryptophan. For quantitative analysis, measurements at a single wavelength, *e.g.* λ_{14}, would be adequate if no interfering species are present. In the presence of other absorbing species, however, more measurements are needed. In Table 10 are presented the concentrations and measured absorbance values at λ_{14} of seven standard

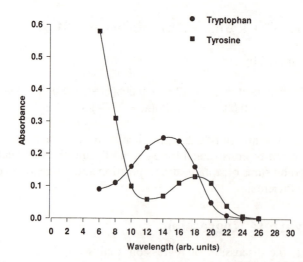

Figure 6   *The UV spectra, recorded at discrete wavelengths, of tryptophan and tyrosine*

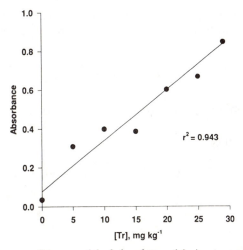

Figure 7 *The least squares linear model of absorbance (A_{14}) vs. concentration of tryptophan, data from Table 10*

Table 10 *Absorbance values of tryptophan standard solutions recorded at two wavelengths, A_{14} and A_{21}. Three 'unknown' test solutions, X1, X2, and X3, are included with their true tryptophan concentration shown in parenthesis*

Tryptophan concn. (mg kg^{-1})	Absorbance A_{14}	Absorbance A_{21}
0	0.0356	0.0390
5	0.3068	0.2110
10	0.3980	0.1860
15	0.3860	0.0450
20	0.6020	0.1580
25	0.6680	0.1070
29	0.8470	0.2010
X1 (7)	0.3440	0.2010
X2 (14)	0.3670	0.0500
X3 (27)	0.0810	0.2110

solutions containing known amounts of tryptophan along with three samples which we will assume contain unknown amounts of tryptophan. All solutions have unknown concentrations of a second absorbing species present, in this case the amino acid tyrosine. The effect of this interferent is to add noise and distort the univariate calibration graph, as shown in Figure 7. The best-fit linear regression line is also shown, as derived from

$$\text{Concentration tryptophan, } Tr = -2.00 + 38.51 A_{14} \qquad (40)$$

where A_i is the absorbance at λ_i.

Despite the apparently high r^2 value for this model ($r^2 = 0.943$), its predictive ability is poor as can be demonstrated with the three test samples:

Actual:	7	14	27	mg kg^{-1}
Predicted:	10.26	11.14	28.20	mg kg^{-1}

If a second term, say the absorbance at λ_{21}, is added to the model equation, the predictive ability is improved considerably. Thus by including A_{21}, the least-squares model is

$$Tr = -0.00067 + 43.68A_{14} - 39.78A_{21} \tag{41}$$

with the test samples,

Actual:	7	14	27	mg kg^{-1}
Predicted:	7.03	14.04	26.99	mg kg^{-1}

This model as given by Equation (41) could be usefully employed for the quantitative determination of tryptophan in the presence of tyrosine.

Of course, the reason for the improvement in the calibration model when the second term is included is that A_{21} serves to compensate for the absorbance due to the tyrosine since λ_{21} is in the spectral region of a tyrosine absorption band with little interference from tryptophan, Figure 6. In general, the selection of variables for multivariate regression analysis may not be so obvious.

Selection of Variables for Regression

In the discussions above and in the examples previously described, it has been assumed that the variables to be included in the multivariate regression equation were known in advance. Either some theoretical considerations determine the variables or, as in many spectroscopic examples, visual inspection of the data provides an intuitive feel for the greater relevance of some variables compared with others. In such cases, serious problems associated with the selection of appropriate variables may not arise. The situation is not so simple where no sound theory exists and variable selection is not obvious. Then some formal procedure for choosing which variables to include in a regression analysis is important and the task may be far from trivial.

The problems and procedures for selecting variables for regression analysis can be illustrated by considering the use of near-IR spectrometry for quantitative analysis. Despite its widespread use in manufacturing and process industries, the underlying theory regarding specific spectral transitions associated with the absorption of radiation in the near-IR region has been little studied. Unlike the fundamental transitions observed in the mid-IR region, giving rise to discrete absorption bands, near-IR spectra are often characterized by overtones and combination bands and the observed spectra are typically complex and, to a large extent, lacking in readily identifiable features. It rarely arises, therefore, that absorption at a specific wavelength can be attributed to a single chemical entity or species. For quantitative analysis a range of measurements, each at different wavelengths, must be recorded in order to attempt to correct for spectral interference. In the limit, of course, the whole spectrum can be employed as a list or vector of variates. The dependent variable, y, can then be represented by a linear model of the form

$$y = a + \Sigma b_i x_i + \epsilon_i \tag{42}$$

where y is the concentration of some analyte, x_i is the measured response (absorbance or reflectance) at i specific wavelengths, and a and b are the coefficients or weights associated with each variate. For a complete spectrum, extending from say 1200 to 2000 nm, i may take on values of several hundreds and the solution of the possible hundreds of simultaneous equations necessary to determine the full range of the coefficients in order to predict y from the analytical data is computationally demanding. In preparing such a multivariate calibration model, therefore, it would be reasonable to address two key points. Firstly, which of the variates contribute most significantly to the prediction model and which variates can be left out without reducing the effectiveness of

Table 11 *UV absorbance data recorded at seven wavelengths, $A_9 \ldots A_{27}$, of 14 solutions containing known amounts of tryptophan. The spectra of two test solutions containing 11 and 25 mg kg^{-1} tryptophan respectively are also included*

Tr (mg kg^{-1})	A_9	A_{12}	A_{15}	A_{18}	A_{21}	A_{24}	A_{27}
2	0.632	0.292	0.318	0.436	0.296	0.069	0.079
4	0.558	0.275	0.418	0.468	0.258	0.116	0.072
6	0.565	0.300	0.392	0.501	0.279	0.040	0.052
8	0.549	0.332	0.502	0.509	0.224	0.055	0.018
10	0.570	0.351	0.449	0.480	0.222	0.056	0.025
12	0.273	0.309	0.427	0.324	0.156	0.056	0.080
14	0.276	0.378	0.420	0.265	0.063	0.019	0.006
16	0.469	0.444	0.550	0.456	0.181	0.063	0.053
18	0.504	0.551	0.585	0.524	0.172	0.110	0.078
20	0.554	0.566	0.654	0.513	0.168	0.070	0.083
22	0.501	0.553	0.667	0.521	0.143	0.103	0.035
24	0.464	0.636	0.691	0.525	0.122	0.077	0.100
26	0.743	0.743	0.901	0.785	0.313	0.088	0.072
28	0.754	0.793	0.939	0.773	0.261	0.024	0.095
Mean 15	0.529	0.466	0.565	0.506	0.204	0.068	0.061
s 8.367	0.139	0.174	0.188	0.139	0.072	0.030	0.030
*X*1 11	0.254	0.324	0.337	0.337	0.110	0.035	0.034
*X*2 25	0.497	0.656	0.771	0.513	0.150	0.053	0.083

Table 12 *Correlation matrix between tryptophan concentration and absorbance at seven wavelengths for the 14 standard solutions from Table 11*

	Tr	A_9	A_{12}	A_{15}	A_{18}	A_{21}	A_{24}	A_{27}
Tr	1							
A_9	0.225	1						
A_{12}	0.955	0.474	1					
A_{15}	0.919	0.554	0.969	1				
A_{18}	0.589	0.877	0.765	0.832	1			
A_{21}	-0.228	0.830	0.020	0.135	0.615	1		
A_{24}	0.015	0.250	0.083	0.108	0.220	0.207	1	
A_{27}	0.393	0.376	0.510	0.474	0.456	0.271	0.298	1

the model? If most of the calibration information can be demonstrated to reside in only a few measurements then the computational effort is reduced considerably. Secondly, is there any penalty, other than increased data processing time, in having more variates in the set of equations than strictly necessary? After all, with the data processing power now available with even the most modest personal computer, why not include all measurements in the calibration?

As an easily managed example of multivariate data analysis we shall consider the spectral data presented in Table 11. These data represent the recorded absorbance of 14 standard solutions containing known amounts of tryptophan, measured at seven wavelengths, in the UV region under noisy conditions and in the presence of other absorbing species. Two test spectra, $X1$ and $X2$, are also included.

Some of these spectra are illustrated in Figure 8 and the variation in absorbance at each wavelength as a function of tryptophan concentration is shown in Figure 9. No single wavelength measure exhibits an obvious linear trend with analyte concentration and a univariate calibration is unlikely to prove successful. The matrix of correlation coefficients between the variables, dependent and independent, is given in Table 12. The independent variable most highly correlated with tryptophan concentration is the measured absorbance at λ_{12}, A_{12}, *i.e.*

$$Tr = a + b_1 A_{12} \tag{43}$$

and by least-squares modelling,

$$Tr = -6.31 + 45.74 A_{12} \tag{44}$$

and for our two test samples, of concentrations 11 and 25 p.p.m., $X1 = 8.51$ mg kg^{-1} and $X2 = 23.69$ mg kg^{-1}.

The regression model *vs.* actual results scatter plot is shown in Figure 10 and the plot of residuals ($y_i - \hat{y}_i$) in Figure 11. Despite the apparent high correlation between tryptophan concentration and A_{12}, the univariate model is a poor predictor, particularly at low concentrations.

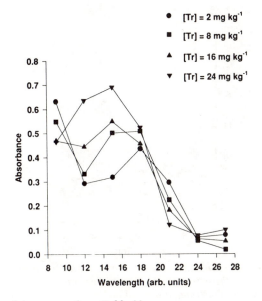

Figure 8 *Some of the spectra from Table 11*

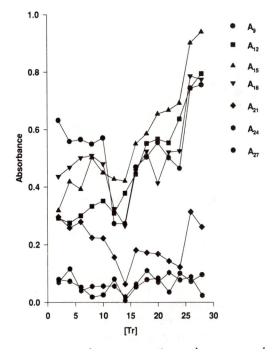

Figure 9 *Absorbance vs. tryptophan concentration at the seven wavelengths monitored*

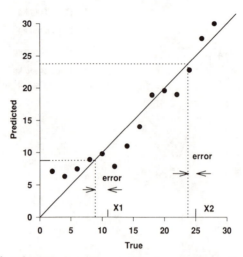

Figure 10 *The predicted tryptophan concentration from the univariate regression model, using A_{12}, vs. the true, known concentration. Prediction lines for test samples X1 and X2 are illustrated also*

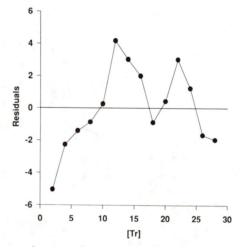

Figure 11 *Residuals as a function of concentration for the univariate regression model, using A_{12} from Table 11*

In order to improve the performance of the calibration model other information from the spectral data could be included. The absorbance at λ_{21}, for example, is negatively correlated with tryptophan concentration and may serve to compensate for the interfering species present. Including A_{21} gives the bivariate model defined by

$$Tr = a + b_1 A_{12} + b_2 A_{21} \tag{45}$$

By ordinary least squares regression, Equation (45) can be solved to provide

$$Tr = -0.51 + 45.75A_{12} - 28.43A_{21} \qquad (46)$$

with a coefficient of determination, r^2, of 0.970. The model *vs.* actual data and the residuals plot are shown in Figures 12 and 13. $X1$ and $X2$ are evaluated as 11.19 and 25.24 mg kg^{-1} respectively.

Although the bivariate model performs considerably better than the uni-variate model, as evidenced by the smaller residuals, the calibration might be improved further by including more spectral data. The question arises as to which data to include. In the limit of course, all data will be used and the model takes the form

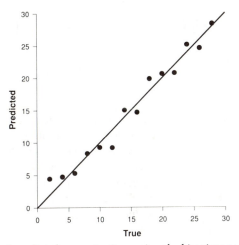

Figure 12   *True and predicted concentrations using the bivariate model with A_{12} and A_{21}*

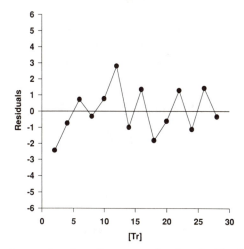

Figure 13   *Residuals as a function of concentration for the bivariate regression model, using A_{12} and A_{21} from Table 11*

$$Tr = a + b_1 A_9 + b_2 A_{12} + b_3 A_{15} + \ldots + b_7 A_{27} \tag{47}$$

To determine by least-squares the value of each coefficient requires we use eight simultaneous equations. In matrix notation the normal equations can be expressed as,

$$Tr = A.b \tag{48}$$

where

$$A = \begin{bmatrix} \Sigma 1 & \Sigma A_9 & \ldots & \Sigma A_{27} \\ \Sigma A_9 & \Sigma A_9{}^2 & & \Sigma A_9 A_{27} \\ \Sigma A_{12} & \Sigma A_{12} A_9 & & \Sigma A_{12} A_{27} \\ \vdots & & & \\ \Sigma A_{27} & & & \Sigma A_{27}{}^2 \end{bmatrix}$$

$$b^{\mathrm{T}} = [a\ b_1\ b_2\ b_3\ b_4\ b_5\ b_6\ b_7]$$
$$Tr^{\mathrm{T}} = [\Sigma Tr\ \Sigma A_9 Tr\ \Sigma A_{12} Tr\ \Sigma A_{15} Tr\ \Sigma A_{18} Tr\ \Sigma A_{21} Tr\ \Sigma A_{24} Tr\ \Sigma A_{27} Tr] \tag{49}$$

Calculating the individual elements of matrix A and computing its inverse in order to solve Equation (48) for b can give rise to computational errors, and it is common practice to modify the calculation to achieve greater accuracy.[5]

If the original data matrix is converted into the correlation matrix, then each variable is expressed in the standard normal form with zero mean and unit standard deviation. The intercept coefficient using these standardized variables will now be zero and the required value can be calculated later. The regression equation in matrix form is then

$$R.B = r \tag{50}$$

where R is the matrix of correlation coefficients between the independent variables, r is the vector of correlations between the dependent variable and each independent variable, and B is the vector of standard regression coefficients we wish to determine.

The individual elements of R and r are available from Table 12 and we may calculate B by rearranging Equation (50):

$$B = R_x^{-1}.r_y \tag{51}$$

and,

$$B^{\mathrm{T}} = [-0.28\ \ 0.709\ \ 0.419\ \ -0.006\ \ -0.052\ \ 0.006\ \ -0.044]$$

(displayed as a row vector for convenience only).

[5] J.C. Davis, 'Statistics in Data Analysis in Geology', J. Wiley and Sons, New York, USA, 1973.

To be used in a predictive equation these coefficients must be 'unstandard-ized', and, from Equation (30),

$$b_i = B_i \sigma_y / \sigma_i$$

Hence

$$\boldsymbol{b}^{\mathrm{T}} = [-16.83 \quad 34.08 \quad 18.67 \quad -0.36 \quad -6.08 \quad 1.71 \quad -12.38] \quad (52)$$

The constant intercept term is obtained from Equation (47),

$$
\begin{aligned}
a &= y - b_1 \bar{x}_1 - b_2 \bar{x}_2 - b_3 \bar{x}_3 - b_4 \bar{x}_4 - b_5 \bar{x}_5 - b_6 \bar{x}_6 - b_7 \bar{x}_7 \\
&= -0.465
\end{aligned} \quad (53)
$$

Predicted regression results compared with known tryptophan concentration values are shown graphically in Figure 14, and Figure 15 shows the residuals. The calculated concentrations for $X1$ and $X2$ are 11.44 and 25.89 p.p.m. respectively. Although the predicted concentrations for our two test samples are inferior to the results obtained with the bivariate model, the full, seven-factor model fits the data better as can be observed from Figure 14 and the smaller residuals in Figure 15. Unfortunately, including all seven terms in the model has also added random noise to the system; A_{24} and A_{27} are measured at long wavelengths where negligible absorption would be expected from any component in the samples. In addition, where several hundred wavelengths may be monitored with a high degree of colinearity between the data, it is necessary and worthwhile using an appropriate subset of the independent variables. For predictive purposes it is often possible to do at least as well with a carefully chosen subset as with the total set of independent variables. As the

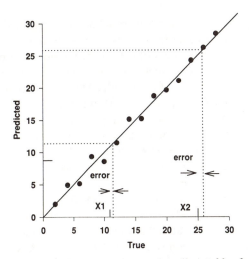

Figure 14 *True and predicted concentrations using all variables from Table 11*

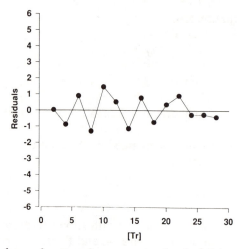

Figure 15   *Residuals as a function of concentration for the full regression model using all variables from Table 11*

number of independent variables increases the number of subsets of all possible combinations of variables increases dramatically and a formal procedure must be implemented to select the most appropriate variables to include in the regression model. A very direct procedure for testing the significance of each variable involves fitting all possible subsets of the variates in the equation and evaluating the best response. However, this is rarely possible. With p variables the total number of equations to be examined is 2^p, if we include the equation containing all variates and that containing none. Even with only eight variables, the number of equations is 256, and to examine a complete spectrum containing many hundreds of measures the technique is neither feasible nor practical.

In some cases there may exist a strong practical or theoretical justification for including certain variables in the regression equation. In general, however, there is no preconceived assessment of the relative importance of some or all of the independent variables. One method, mentioned briefly previously, is to examine the relative magnitudes of the standard regression coefficients. For our experimental data, from $\boldsymbol{B}^T$ [Equation (51)], this would indicate that A_9, A_{12}, and A_{15} are the most important. More sophisticated strategies are employed in computer software packages. For cases where there are a large number of significant variates, three basic procedures are in common use. These methods are referred to as the *forward selection procedure*, the *backward selection procedure*, and the *stepwise method*.

The forward selection technique starts with an empty equation, possibly containing a constant term only, with no independent variables. As the procedure progresses, variates are added to the test equation one at a time. The first variable included is that which has the highest correlation with the dependent variable y. The second variable added to the equation is the one with the highest correlation with y, after y has been adjusted for the effect of the first variable, *i.e.* the variable with the highest correlation with the residuals from the first step. This method is equivalent to selecting the second variable so as to maximize the partial correlation with y after removing the linear effect of the first chosen variable. The procedure proceeds in this manner until no further variate has a significant effect on the fitted equation.

From Table 12, the absorbance at λ_{12} exhibits the highest correlation with tryptophan concentration and this is the first variable added to the equation, Equation (43). To choose the second variable, we could select A_{15} as this has the second highest absolute correlation with Tr but this may not be the best choice. Some other variable combined with A_{12} may give a higher *multiple correlation* than A_{15} and A_{12}.

Multiple correlation represents the simple correlation between known values of the dependent variable and equivalent points or values as derived from the regression equation. *Partial correlation*, on the other hand, is the simple correlation between the residuals from the regression line or planes on the variable whose effects are removed.[6] For our UV absorbance data we wish to remove the linear effect of A_{12} regressed on Tr so that we can subsequently assess the correlations of the other variables.

From Equation (44), for the univariate model using A_{12},

$$Tr = -6.31 + 45.74A_{12}$$

and regressing A_{12} on to each of the remaining independent variables gives

$$\begin{aligned}
A_9 &= 0.32 + 0.40A_{12} \\
A_{15} &= 0.06 + 1.07A_{12} \\
A_{18} &= 0.21 + 0.60A_{12} \\
A_{21} &= 0.19 + 0.08A_{12} \\
A_{24} &= 0.06 + 0.01A_{12} \\
A_{27} &= 0.02 + 0.08A_{12}
\end{aligned} \tag{54}$$

The matrix of residuals ($Tr - \hat{Tr}$, $A_9 - \hat{A}_9$, $A_{15} - \hat{A}_{15}$, *etc.*) is given in Table 13, and the corresponding correlation matrix between these residuals in Table 14. From Table 14 the variable having the largest absolute correlation with Tr residuals is A_9. Therefore we select this as the second variable to be added to the regression model.

Hence, at step 2,

$$Tr = -0.60 + 52.10A_{12} - 16.58A_9 \tag{55}$$

Forward regression proceeds to step 3 using the same technique. The variables A_9 and A_{12} are regressed on to each of the variables not in the equation

---

[6] A.A. Afifi and V. Clark, 'Computer-Aided Multivariate Analysis', Lifetime Learning, California, USA, 1984.

Table 13 *Matrix of residuals for each variable after removing the linear model using A_{12}*

$Tr - \hat{Tr}$	$A_9 - \hat{A}_9$	$A_{15} - \hat{A}_{15}$	$A_{18} - \hat{A}_{18}$	$A_{21} - \hat{A}_{21}$	$A_{24} - \hat{A}_{24}$	$A_{27} - \hat{A}_{27}$
− 5.330	0.205	− 0.054	0.051	0.083	0.005	0.036
− 2.557	0.138	0.067	0.093	0.046	0.052	0.030
− 1.694	0.135	0.011	0.111	0.065	− 0.024	0.008
− 1.149	0.106	0.087	0.100	0.007	− 0.010	− 0.029
− 0.013	0.120	0.013	0.059	0.004	− 0.009	− 0.023
3.897	− 0.161	0.036	− 0.071	− 0.059	− 0.008	0.035
2.759	− 0.185	− 0.044	− 0.172	− 0.157	− 0.046	− 0.044
1.757	− 0.019	0.015	− 0.020	− 0.044	− 0.003	− 0.002
− 1.109	− 0.026	− 0.065	− 0.017	− 0.062	0.042	0.014
0.208	0.018	− 0.012	− 0.037	− 0.067	0.002	0.018
2.800	− 0.030	0.015	− 0.021	− 0.091	0.035	− 0.029
1.025	− 0.100	− 0.049	− 0.067	− 0.119	0.008	0.029
− 1.842	0.136	0.046	0.129	0.064	0.018	− 0.007
− 2.116	0.127	0.031	0.087	0.008	− 0.047	0.012

Table 14 *Matrix of correlations between the residuals from Table 13*

	$Tr - \hat{Tr}$	$A_9 - \hat{A}_9$	$A_{15} - \hat{A}_{15}$	$A_{18} - \hat{A}_{18}$	$A_{21} - \hat{A}_{21}$	$A_{24} - \hat{A}_{24}$	$A_{27} - \hat{A}_{27}$
$Tr - \hat{Tr}$	1						
$A_9 - \hat{A}_9$	− 0.88	1					
$A_{15} - \hat{A}_{15}$	0.08	0.34	1				
$A_{18} - \hat{A}_{18}$	− 0.73	0.91	0.57	1			
$A_{21} - \hat{A}_{21}$	− 0.81	0.92	0.42	0.91	1		
$A_{24} - \hat{A}_{24}$	− 0.15	0.12	0.01	0.16	0.11	1	
$A_{27} - \hat{A}_{27}$	− 0.35	0.15	− 0.18	0.11	0.29	0.29	1

and the unused variable with the highest partial correlation coefficient is selected as the next to use. If we continue in this way then all variables will eventually be added and no effective subset will have been generated, so a *stopping rule* is employed. The most commonly used stopping rule in commercial programs is based on the F-test of the hypothesis that the partial correlation coefficient of the variable to be entered in the equation is equal to zero. No more variables are added to the equation when the F-value is less than some specified cut-off value, referred to as the *minimum F-to-enter* value.

A completed forward regression analysis of the UV absorbance data is presented in Table 15. Using a cut-off F-value of 4.60 ($F_{1,14}$ at 95% confidence limit), three variables are included in the final equation:

$$Tr = -0.77 + 33.92A_{12} - 19.47A_9 + 18.05A_{15} \tag{56}$$

The predicted *vs.* actual data are illustrated in Figure 16 and the residuals

Table 15 *Forward regression analysis of the data from Table 11. After three steps no remaining variable has a F-to-enter value exceeding the declared minimum of 4.60, and the procedure stops*

Step 1: Variable entered: A_{12}

Dependent variable		Tr				
Variables in equation	Constant	A_{12}				
Coefficient	−6.31	45.74				
Variables not in equation	A_9	A_{15}	A_{18}	A_{21}	A_{24}	A_{27}
Partial correlation coefficient	−0.88	−0.10	−0.74	−0.83	−0.22	−0.36
F-to-enter	43.20	0.14	16.19	29.87	0.64	1.90

Step 2: Variable entered: A_9

Variables in equation	Constant	A_{12}		A_9	
Coefficient	−0.60	52.10		−16.58	
Variables not in equation	A_{15}	A_{18}	A_{21}	A_{24}	A_{27}
Partial correlation coefficient	0.65	0.25	−0.10	−0.01	−0.43
F-to-enter	8.84	0.87	0.12	0.03	2.66

Step 3: Variable entered: A_{15}

Variables in equation	Constant	A_{12}	A_9	A_{15}
Coefficient	−0.77	33.92	−19.47	18.05
Variables not in equation	A_{18}	A_{21}	A_{24}	A_{27}
Partial correlation coefficient	−0.07	−0.30	−0.03	−0.40
F-to-enter	0.06	1.10	0.01	2.11

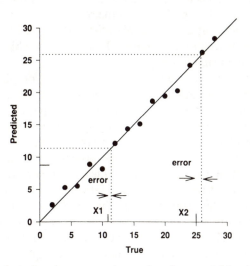

Figure 16 *True and predicted concentrations using three variables (A_9, A_{12}, and A_{21}) from Table 11*

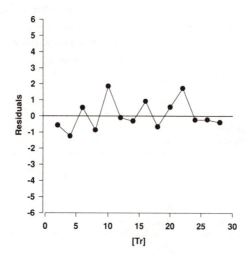

Figure 17   *Residuals as a function of concentration for three variable regression model from forward regression analysis*

plotted in Figure 17. Calculated values for $X1$ and $X2$ are 11.30 and 25.72 mg kg^{-1} respectively.

An alternative method is described by backward elimination. This technique starts with a full equation containing every measured variate and successively deletes one variable at each step. The variables are dropped from the equation on the basis of testing the significance of the regression coefficients, *i.e.* for each variable is the coefficient zero? The F-statistic is referred to as the computed *F-to-remove*. The procedure is terminated when all variables remaining in the model are considered significant.

Table 16 illustrates a worked example using the tryptophan data. Initially, with all variables in the model, A_{18} has the smallest computed F-to-remove value and this variable is removed from the model and eliminated at the first step. The procedure proceeds by computing a new regression equation with the remaining six variables and again examining the calculated F-to-remove values for the next candidate for elimination. This process continues until no variable can be removed since all F-to-remove values are greater than some specified maximum value. This is the stopping rule; F-to-remove = 4 was employed here.

It so happens in this example that the results of performing backward elimination regression are identical with those obtained from the forward regression analysis. This may not be the case in general. In its favour, forward regression generally involves a smaller amount of computation than backward elimination, particularly when many variables are involved in the analysis. However, should it occur that two or more variables combine together to be a good predictor compared with single variables, then backward elimination will often lead to a better equation.

Finally, stepwise regression, a modified version of the forward selection technique, is often available with commercial programs. As with forward

Table 16 *Backward regression analysis of the data from Table 11. After four steps, three variables remain in the regression equation; their F-to-remove values exceed the declared maximum value of 4.0*

Step 0: All variables entered
Dependent variable Tr
Variables

in equation	Constant	A_9	A_{12}	A_{15}	A_{18}	A_{21}	A_{24}	A_{27}
Coefficient	− 0.465	− 16.83	34.08	18.67	− 0.36	− 6.08	1.71	− 12.38
F-to-Remove		5.96	12.43	5.03	0.001	0.08	0.04	0.78

Step 1: Remove A_{18}
Variables

in equation	Constant	A_9	A_{12}	A_{15}	A_{21}	A_{24}	A_{27}
Coefficient	− 0.62	− 16.45	34.85	17.12	− 4.83	2.09	− 13.25
F-to-Remove		6.92	16.54	6.32	0.17	0.04	1.01

Step 2: Remove A_{24}

Variables in equation	Constant	A_9	A_{12}	A_{15}	A_{21}	A_{27}
Coefficient	− 0.53	− 16.14	34.50	17.28	− 5.32	− 12.35
F-to-Remove		7.83	18.67	7.23	0.23	1.09

Step 3: Remove A_{21}

Variables in equation	Constant	A_9	A_{12}	A_{15}	A_{27}
Coefficient	− 0.62	− 18.69	36.67	16.37	− 14.88
F-to-Remove		73.03	33.29	7.64	2.12

Step 4: Remove A_{27}

Variables in equation	Constant	A_9	A_{12}	A_{15}
Coefficient	− 0.77	− 19.47	33.92	18.05
F-to-Remove		77.08	25.89	8.84

selection, the procedure increases the number of variables in the equation at each step but at each stage the possibility of deleting a previously included variable is considered. Thus, a variable entered at an earlier stage of selection may be deleted at subsequent, later stages.

It is important to bear in mind that none of these subset multiple linear regression techniques are guaranteed or even expected to produce the best possible regression equation. The user of commercial software products is encouraged to experiment.

Principal Components Regression

It is often the case with multiple regression analysis involving large numbers of independent variables that there exists extensive colinearity or correlation between these variables. Colinearity adds redundancy to the regression model since more variables may be included in the model than is necessary for adequate predictive performance. Of the methods available to the analytical chemist for regression analysis with protection against the problems induced by correlation between variables, principal components regression, PCR, is the most common employed.

Having discussed in the previous section the problems associated with variable selection, we may now summarize our findings. The following rules-of-thumb provide a useful guide:

(a) Select the smallest number of variables possible. Including unnecessary variables in our model will introduce bias in the estimation of the regression coefficients and reduce the precision of the predicted values.
(b) Use the maximum information contained in the independent variables. Although some of the variables are likely to be redundant, potentially important variables should not be discounted solely in order to reduce the size of the problem.
(c) Choose independent variables that are not highly correlated with each other. Colinearity can cause numerical instability in estimating regression coefficients.

Although subset selection along with multiple linear regression provides a means of reducing the number of variables studied, the method does not address the problems associated with colinearity. To achieve this, the regression coefficients should be orthogonal. The technique of generating orthogonal linear combinations of variables in order to extract maximum information from a data set was encountered previously in eigen analysis and the calculation of principal components. The ideas derived and developed in Chapter 3 can be applied here to regression analysis.

As an example consider the variables A_{12}, A_{15}, and A_{18} from the UV absorbance data of Table 11. These three variables are highly correlated between each other as can be seen from Table 12. This intercorrelation can also be observed in the scatter plot of these variables, Figure 18. By principal

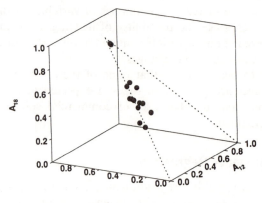

Figure 18 *Scatter plot of absorbance data at three wavelengths, A_{12}, A_{15}, and A_{18}, from Table 11. The high degree of colinearity, or correlation, between these data is evidenced by their lying on a plane and not being randomly distributed in the pattern space*

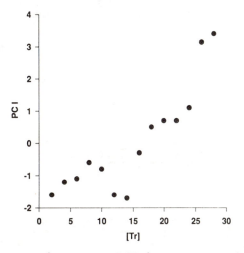

Figure 19 *The first principal component, PCI, from A_{12}, A_{15}, and A_{18} vs. tryptophan concentration*

components analysis two new variables can be defined containing over 99% of the original variance of the three original variables. The first principal component alone accounts for over 90% of the total variance and a plot of *Tr* against PCI is shown in Figure 19.

The use and application of principal components in regression analysis has been extensively reported in the chemometrics literature.[7-10] We can calculate the principal components from our data set, so providing us with a set of new, orthogonal variables. Each of these principal components will be a linear combination of, and contain information from, all the original variables. By selecting an appropriate subset of principal components, the regression model is reduced whilst having the relevant information from the original data. The PCR technique described here follows the methodology described by Martens and Naes[11] and is applied to the data from Table 11. The original data are preprocessed by mean-centring. The variance–covariance dispersion matrix is then computed and from this square, symmetric matrix we calculate the normalized eigenvalues and eigenvectors. From each eigenvector, the principal component scores are determined, and by conventional regression analysis the calibration model is developed. The stepwise procedure is illustrated in Table 17 and we will now follow the steps involved.

[7] E.V. Thomas and D.M. Haaland, *Anal. Chem.*, 1990, **62**, 1091.
[8] R.G. Brereton, 'Chemometrics', Ellis Horwood, Chichester, UK, 1990.
[9] J.H. Kalivas, in 'Practical Guide to Chemometrics', ed. S.J. Haswell, Marcel Dekker, New York, USA, 1992.
[10] P.S. Wilson, in 'Computer Methods in UV, Visible and IR Spectroscopy', ed. W.O. George and H.A. Willis, Royal Society of Chemistry, Cambridge, UK, 1990.
[11] H. Martens and T. Naes, 'Multivariate Calibration', J. Wiley and Sons, Chichester, UK, 1991.

Table 17 *Principal components regression[11] on the data from Table 11. The steps involved are discussed in the text*

	A_9	A_{12}	A_{15}	A_{18}	A_{21}	A_{24}	A_{27}	Tr	t_1
Mean	0.529	0.466	0.565	0.506	0.204	0.068	0.061	15.00	
				X_0				y_0	
	0.103	−0.174	−0.247	−0.070	0.092	0.011	0.018	−13	0.237
	0.029	−0.191	−0.147	−0.038	0.054	0.048	0.011	−11	0.198
	0.036	−0.166	−0.173	−0.005	0.075	−0.028	−0.009	−9	0.184
	0.020	−0.134	−0.063	0.003	0.020	0.013	−0.043	−7	0.106
	0.041	−0.115	−0.116	−0.026	0.018	−0.012	−0.036	−5	0.133
	−0.256	−0.157	−0.138	−0.182	−0.048	−0.012	0.019	−3	0.341
	−0.253	−0.088	−0.145	−0.241	−0.141	−0.049	−0.055	−1	0.343
	−0.060	−0.022	−0.015	−0.050	−0.023	−0.005	−0.008	1	0.066
	−0.025	0.085	0.020	0.018	−0.032	0.042	0.017	3	−0.058
	0.025	0.100	0.089	0.007	−0.036	0.002	0.022	5	−0.120
	−0.028	0.087	0.102	0.015	−0.061	0.035	−0.026	7	−0.103
	−0.065	0.170	0.126	0.019	−0.082	0.009	0.039	9	−0.154
	0.214	0.277	0.336	0.279	0.109	0.020	0.011	11	−0.564
	0.225	0.327	0.374	0.267	0.057	−0.044	0.034	13	−0.610
p_1	−0.334	−0.555	−0.614	−0.442	−0.076	−0.010	−0.045	$q_1 =$	−23.405
b	7.82	12.99	14.37	10.34	1.78	0.24	1.06	$a =$	−8.99

	A_9	A_{12}	A_{15}	A_{18} X_1	A_{21}	A_{24}	A_{27}	t_2
	0.182	-0.042	-0.102	0.035	0.110	0.004	0.029	-0.231
	0.095	-0.081	-0.025	0.050	0.069	0.050	0.020	-0.151
	0.097	-0.064	-0.060	0.076	0.089	-0.026	-0.001	-0.172
	0.055	-0.075	0.002	0.050	0.028	-0.011	-0.038	-0.093
	0.085	-0.041	-0.034	0.033	0.028	-0.010	0.030	-0.106
	-0.143	0.032	0.071	-0.031	-0.022	-0.008	0.035	0.149
	-0.139	0.102	0.065	-0.089	-0.115	-0.045	-0.039	0.235
	-0.038	0.015	0.025	-0.021	-0.018	-0.004	-0.005	0.054
	-0.045	0.053	-0.016	-0.007	-0.037	0.042	0.015	0.064
	-0.015	0.034	0.015	-0.046	-0.045	0.001	0.017	0.063
	-0.063	0.030	0.039	-0.030	-0.069	0.034	-0.030	0.106
	-0.117	0.085	0.031	-0.049	-0.094	0.008	0.032	0.179
	0.025	-0.036	0.010	0.030	0.066	0.015	-0.014	-0.074
	0.021	-0.011	-0.001	-0.002	0.010	0.050	0.007	-0.022
p_2	-0.672	0.383	0.300	-0.308	-0.465	-0.023	-0.007	
b	13.96	25.40	24.09	0.35	-13.29	-0.52	0.83	$a = -0.54$

Table 17 *continued*

	A_9	A_{12}	A_{15}	A_{18} X_2	A_{21}	A_{24}	A_{27}	t_3
	0.026	0.046	− 0.032	− 0.036	0.002	− 0.002	0.028	0.073
	− 0.007	− 0.023	0.020	0.003	− 0.001	0.047	0.019	− 0.015
	− 0.019	0.002	− 0.009	0.023	0.009	− 0.030	− 0.001	− 0.016
	− 0.008	− 0.040	0.030	0.021	− 0.015	− 0.014	− 0.038	− 0.057
	0.014	− 0.001	− 0.003	0.000	− 0.021	− 0.013	− 0.030	0.001
	− 0.042	− 0.025	0.026	0.015	0.047	− 0.005	0.036	− 0.050
	0.019	0.013	− 0.005	− 0.017	− 0.006	− 0.040	− 0.037	0.009
	− 0.002	− 0.006	0.009	− 0.004	0.007	− 0.003	− 0.004	− 0.011
	− 0.002	0.029	− 0.035	0.012	− 0.007	0.043	0.015	0.040
	0.027	0.010	− 0.004	− 0.026	− 0.016	0.003	0.017	0.036
	0.009	− 0.011	0.007	0.003	− 0.020	0.037	− 0.029	− 0.007
	0.003	0.016	− 0.022	0.006	− 0.011	0.012	0.034	0.033
	− 0.025	− 0.008	0.012	0.007	0.032	0.013	− 0.015	− 0.032
	0.006	0.003	0.006	− 0.009	0.000	− 0.050	0.007	− 0.004
p_3	0.401	0.578	− 0.499	− 0.331	− 0.218	0.126	0.288	
b	− 12.95	26.86	22.83	− 0.49	− 13.84	− 0.20	1.56	$a = $ − 0.571

A_9	A_{12}	A_{15}	A_{18} X_3	A_{21}	A_{24}	A_{27}
-0.003	0.004	0.004	-0.012	0.018	-0.011	0.007
-0.000	-0.014	0.012	-0.002	-0.005	0.049	0.024
-0.012	0.011	-0.017	0.018	0.005	-0.028	0.003
0.015	-0.007	0.001	0.002	-0.028	-0.006	-0.022
0.014	-0.001	-0.002	0.001	-0.021	-0.013	-0.031
-0.022	0.004	0.001	-0.002	0.036	0.002	0.050
0.015	0.008	-0.001	-0.014	-0.004	-0.041	-0.040
0.002	0.000	0.004	-0.008	0.005	-0.001	-0.001
-0.018	0.005	-0.015	0.026	0.002	0.038	0.004
0.012	0.011	0.014	0.014	0.008	0.002	0.007
0.011	-0.007	0.004	0.001	-0.021	0.038	-0.028
-0.010	-0.003	-0.006	0.017	-0.004	0.008	0.024
-0.012	0.011	-0.004	-0.003	0.025	0.017	-0.005
0.008	-0.001	0.004	-0.010	-0.001	-0.050	0.008
P_4 0.401	0.578	-0.499	-0.331	-0.218	0.126	0.288

The general linear regression equation is given by

$$y = a + X.b \tag{57}$$

where b is the vector of estimates of the regression coefficients to be determined. This vector b can be written as the product of the eigenvectors and the y-loadings,

$$b = P.q \tag{58}$$

where P is the matrix of eigenvectors, or loadings. Each column of P is an eigenvector for each factor included in the regression model. Elements of the P matrix are p_{jk} ($j = 1 \ldots m$, the number of original variables, and $k = 1 \ldots K$, the number of factors or principal components used in the model). The vector q represents the y-loadings which can be determined by regression of y on T, the matrix of scores, t_k for each principal component. Martens and Naes derive q from

$$q = D.T^{\mathrm{T}}.y \tag{59}$$

where D is a diagonal matrix, with each diagonal element equal to $1/\tau_k$ ($\tau_k =$ the eigenvalue of factor k).

Working through our example in Table 17 will serve to illustrate the technique in operation. The mean-centred transformed data ($x_i - \bar{x}$ and $y_i - \bar{y}$) is presented as the matrix X_0 and vector y_0. The eigenvalues of X_0 are provided in Table 18 and it is evident that the data can be adequately described by two or three principal components. The eigenvector corresponding to the first principal component is p_1,

$$p_1^{\mathrm{T}} = [-0.334 \quad -0.555 \quad -0.614 \quad -0.442 \quad -0.076 \quad -0.01 \quad -0.045] \tag{60}$$

and the greatest weights are given to A_9, A_{12}, A_{15}, and A_{18}.

The vector of scores, t_1, is obtained from,

$$t_1 = X_0.p_1$$

With only one principal component in the model, $T = t_1$ and $D = 1/\tau_1 = 0.879$; thus,

$$q_1 = 0.879 t_1 y_0 = -23.405 \tag{61}$$

and the estimated regression coefficients for the one-factor model, from Equation (58), are

$$b = p_1.q_1 \tag{62}$$

Table 18 *Eigenvalues of the mean-centred original data (Table 11). Over 97% of the original variance can be accounted for in the first two principal components*

Eigenvalue	Percentage contribution	Cumulative percentage contribution
1.138	79.08	79.08
0.263	18.28	97.36
0.017	1.18	98.54
0.014	0.97	99.51
0.003	0.21	99.72
0.002	0.14	99.86
0.002	0.14	100.00

with a constant a given by

$$a = y - x^{\mathrm{T}}.b \tag{63}$$

Therefore,

$$Tr(\text{single factor}) = -8.99 + 7.82A_9 + 12.99A_{12} + 14.37A_{15}$$
$$+ 10.34A_{18} + 11.78A_{21} + 0.24A_{24} + 1.06A_{27} \tag{64}$$

The actual *vs.* predicted tryptophan concentration values employing the single factor are shown in Figure 20.

The algorithm proceeds by subtracting the effect of the first factor from X_0

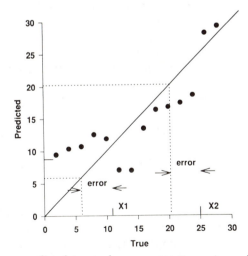

Figure 20 *True vs. predicted tryptophan concentration using only the first principal component in the regression model*

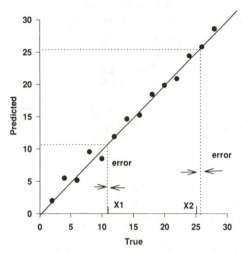

Figure 21 *True vs. predicted tryptophan concentration using the first two principal components in the regression model*

and y_0, yielding the residual matrix X_1 and vector y_1. The process is repeated with the second scores determined from the second eigenvector and a two-factor model developed. Figure 21 shows comparative results.

$$Tr\text{(two factor)} = -0.54 - 13.96A_9 + 25.40A_{12} + 24.09A_{15}$$
$$+ 0.35A_{18} - 13.29A_{21} - 0.52A_{24} + 0.83A_{27} \quad (65)$$

and if a further factor is included,

$$Tr\text{(three factor)} = -0.571 - 12.95A_9 + 26.86A_{12} + 22.83A_{15}$$
$$- 0.49A_{18} - 13.84A_{21} - 0.20A_{24} + 1.56A_{27} \quad (66)$$

with predicted *vs.* actual data shown in Figure 22.

For the one, two, and three factor models the sums of the squares of the y residuals are 337, 11.6, and 11.6 respectively and the predicted concentrations for the test samples X_1 and X_2 are,

	X_1 (11 mg kg^{-1})	X_2 (25 mg kg^{-1})
One factor	5.78	20.18
Two factors	10.93	25.98
Three factors	10.90	26.00

Thus, as anticipated from visual examination of the eigenvalues, two factors are sufficient to describe the calibration and the regression model.

In employing principal components as our regression factors we have succeeded in fully utilizing all the measured variables and developed new, uncorrelated variables. In selecting which eigenvectors to use, the first employed

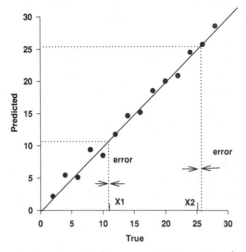

Figure 22 *True vs. predicted tryptophan concentration using the first three principal components in the regression model*

is that corresponding to the largest eigenvalue, the second that corresponding to the next largest eigenvalue, and so on. This strategy assumes that the major eigenvectors correspond to phenomena in the X data matrix of relevance in modelling the dependent variable y. Although this is generally accepted as being the case for most analytical applications, another data compression method can be employed if variables having high variance but little relevance to y are thought to be present. This next method is partial least squares regression.

Partial Least Squares Regression

The calibration model referred to a partial least squares regression (PLSR) is a relatively modern technique, developed and popularized in analytical science by Wold.[12,13] The method differs from PCR by including the dependent variable in the data compression and decomposition operations, *i.e.* both y and x data are actively used in the data analysis. This action serves to minimize the potential effects of x variables having large variances but which are irrelevant to the calibration model. The simultaneous use of X and y information makes the method more complex than PCR as two loading vectors are required to provide orthogonality of the factors.

The method illustrated here employs the orthogonalized PLSR algorithm developed by Wold and extensively discussed by Martens and Naes.[11]

As with PCR, the dependent and independent variables are mean centred to

[12] H. Wold, in 'Perspectives in Probability and Statistics', ed. J. Gani, Academic Press, London, UK, 1975.
[13] H. Wold, in 'Encyclopaedia of Statistical Sciences', ed. N.L. Johnson and S. Kotz, J. Wiley and Sons, New York, USA, 1984.

give data matrix X_0 and vector y_0. Then for each factor, $k = 1 \ldots K$, to be included in the regression model, the following steps are performed.

(a) The loading weight vector w_k is calculated by maximizing the covariance between the linear combination of X_{k-1} and y_{k-1} given that $w_k^T . w_k = 1$.
(b) The factor scores, t, are estimated by projecting X_{k-1} on w_k.
(c) The loading vector p_k is determined by regressing X_{k-1} on t_k and similarly q_k by regressing y_{k-1} on t_k.
(d) From $(X_{k-1} - t_k . p_k^T)$ and $(y_{k-1} - t_k . q_k^T)$ new matrices X_k and y_k are formed.

The optimum number of factors to include in the model is found by observation and usual validation statistics.

For our tryptophan UV absorbance data these steps provide the results shown in Table 19.

The loading weight vector, w_1, is calculated from

$$w_1 = c . X_0 . y_0 \tag{67}$$

where the scaling factor c is given by

$$c = (y_0^T . X_0 . X_0^T . y_0)^{0.5}$$
$$= 0.036 \tag{68}$$

The factor scores and loadings are estimated by

$$\hat{t}_1 = X_0 . \hat{w}_1$$

$$p_1 = (X_0^T . \hat{t}_1)/(\hat{t}_1^T . \hat{t}_1)$$

$$q_1 = (y_0^T . \hat{t}_1)/(\hat{t}_1^T . \hat{t}_1) \tag{69}$$

The matrix and vector of residuals are finally computed,

$$X_1 = X_0 - t_1 . p_1^T \tag{70}$$

and

$$y_1 = y_0 - t_1 . q_1^T \tag{71}$$

The regression coefficients can be calculated by

$$b = W(P^T . W)^{-1} q \tag{72}$$

and

$$a = y - X^T . b \tag{73}$$

Table 19 *Partial least squares regression[11] on the data from Table 11. The steps involved are discussed in the text*

	A_9	A_{12}	A_{15}	A_{18}	A_{21}	A_{24}	A_{27}	Tr	t_1
Mean	0.529	0.466	0.565	0.506 X_0	0.204	0.068	0.061	15.00 y_0	
	0.103	−0.174	−0.247	−0.070	0.092	0.001	0.018	−13	−0.296
	0.029	−0.191	−0.147	−0.038	0.054	0.048	0.011	−11	−0.235
	0.036	−0.166	−0.173	−0.005	0.075	−0.028	−0.009	−9	−0.227
	0.020	−0.134	−0.063	0.003	0.020	0.013	−0.043	−7	−0.129
	0.041	−0.115	−0.116	−0.026	0.018	−0.012	−0.036	−5	−0.159
	−0.256	−0.157	−0.138	−0.182	−0.048	−0.012	0.019	−3	−0.279
	−0.253	−0.088	−0.145	−0.241	−0.141	−0.049	−0.055	−1	−0.255
	−0.060	−0.022	−0.015	−0.050	−0.023	−0.005	−0.008	1	−0.046
	−0.025	0.085	0.020	0.018	−0.032	0.042	0.017	3	0.074
	0.025	0.100	0.089	0.007	−0.036	0.002	0.022	5	0.133
	−0.028	0.087	0.102	0.015	−0.061	0.035	−0.026	7	0.130
	−0.065	0.170	0.126	0.019	−0.082	0.009	0.039	9	0.201
	0.214	0.277	0.336	0.279	0.109	0.020	0.011	11	0.514
	0.225	0.327	0.374	0.267	0.057	−0.044	0.034	13	0.574
w_1^T	0.114	0.647	0.675	0.326	−0.069	0.002	0.039	$q_1 = 26.47$	
p_1^T	0.292	0.599	0.652	0.430	0.043	0.009	0.046	$a = -8.73$	
b	3.012	17.13	17.85	8.64	−1.83	0.05	1.03		

Table 19 *continued*

	A_9	A_{12}	A_{15}	A_{18} X_1	A_{21}	A_{24}	A_{27}	Tr y_1	t_2
	0.189	0.003	− 0.054	0.058	0.105	0.004	0.032	− 5.164	− 0.218
	0.097	− 0.050	0.006	0.063	0.064	0.050	0.022	− 4.776	− 0.141
	0.102	− 0.030	− 0.025	0.093	0.085	− 0.026	0.002	− 2.984	− 0.162
	0.057	− 0.057	0.021	0.059	0.025	− 0.011	− 0.037	− 3.583	− 0.088
	0.087	− 0.020	− 0.012	0.043	0.025	− 0.010	− 0.028	− 0.786	− 0.098
	− 0.175	0.010	0.044	− 0.062	− 0.036	− 0.009	0.032	4.391	0.178
	− 0.179	0.065	0.012	− 0.131	− 0.130	− 0.046	− 0.043	5.742	0.267
	− 0.047	0.006	0.015	− 0.030	− 0.021	− 0.004	− 0.005	2.224	0.060
	− 0.047	0.040	− 0.029	− 0.014	− 0.035	0.042	0.014	1.029	0.061
	− 0.014	0.020	0.002	− 0.050	− 0.042	0.001	0.016	1.474	0.055
	− 0.066	0.009	0.017	− 0.041	− 0.067	0.034	− 0.032	3.557	0.101
	− 0.124	− 0.050	− 0.005	− 0.067	− 0.091	0.008	0.030	3.679	0.171
	0.064	− 0.031	0.000	0.058	0.087	0.016	− 0.012	− 2.612	− 0.119
	0.057	− 0.017	− 0.001	0.021	0.032	− 0.049	0.008	− 2.192	− 0.068
w_2^{T}	− 7.40	0.203	0.092	− 0.430	− 0.466	− 0.029	− 0.028	$q_2 = 23.74$	
p_2^{T}	− 7.42	0.195	0.098	− 0.428	− 0.466	− 0.026	− 0.021	$a = − 0.526$	
b	− 13.90	25.64	23.89	0.30	13.27	− 0.62	0.59		

A_9	A_{12}	A_{15}	A_{18} X_2	A_{21}	A_{24}	A_{27}	Tr y_2	t_3
0.027	0.046	-0.033	-0.036	0.003	-0.002	0.027	0.022	0.034
-0.007	-0.023	0.020	0.003	-0.002	0.047	0.019	-1.441	-0.047
-0.019	0.002	-0.009	0.023	0.009	-0.030	-0.002	0.870	0.008
-0.008	-0.040	0.030	0.021	-0.015	-0.014	-0.038	-1.504	-0.015
0.014	-0.000	-0.003	0.001	0.021	-0.013	-0.030	1.546	0.023
-0.043	-0.025	0.026	0.015	0.047	-0.005	0.036	0.158	-0.052
0.019	0.012	-0.005	-0.017	-0.006	-0.040	0.037	-0.599	0.045
-0.002	-0.006	0.009	-0.004	0.007	-0.003	-0.004	0.795	-0.004
-0.002	0.029	-0.035	0.013	-0.007	0.043	0.015	-0.423	0.012
0.027	0.010	-0.004	-0.026	-0.016	0.003	0.017	0.161	0.002
0.009	-0.011	0.007	0.003	-0.020	0.037	-0.029	0.161	-0.001
0.003	0.016	-0.022	0.006	-0.011	0.012	0.034	-0.389	-0.004
-0.025	-0.008	0.012	0.007	0.031	-0.015	0.210	-0.007	
0.006	-0.003	0.006	-0.009	0.000	-0.050	0.007	-0.567	0.005
w_3^T 0.153	0.572	-0.482	-0.107	0.059	-0.235	-0.589	$q_3 = 9.348$	
p_3^T 0.529	0.594	-0.511	-0.330	-0.329	-0.435	0.364	$a = -0.310$	
b -12.55	31.03	19.41	0.74	-12.78	-2.82	4.92		

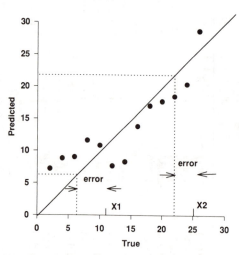

Figure 23 *True vs. predicted tryptophan concentration using a one-factor partial least squares regression model*

where W is the matrix of loading weights, each column is a weight vector, and P the matrix of loadings.

With the single factor in the model the regression equation is,

$$Tr(\text{one factor}) = -8.73 + 3.01A_9 + 17.13A_{12} + 17.85A_{15} \\ + 8.64A_{18} - 1.83A_{21} + 0.05A_{24} + 1.03A_{27} \tag{74}$$

The predicted *vs.* actual concentration as a scatter plot is illustrated in Figure 23.

The procedure is repeated with a second factor included and,

$$Tr(\text{two factors}) = -0.53 - 13.90A_9 + 25.64A_{12} + 23.89A_{15} \\ + 0.30A_{18} - 13.27A_{21} - 0.62A_{24} + 0.59A_{27} \tag{75}$$

and with three factors,

$$Tr(\text{three factors}) = -0.31 - 12.55A_9 + 31.03A_{12} + 19.41A_{15} \\ - 0.74A_{18} - 12.78A_{21} - 2.82A_{24} - 4.92A_{27} \tag{76}$$

The scatter plots are shown in Figure 24. The sums of squares of the residuals for the one, two, and three-factor models are 201, 11.53, and 10.64 respectively and the estimated tryptophan concentrations from the test solutions are

	X_1 (11 mg kg^{-1})	X_2 (25 mg kg^{-1})
One factor	6.35	22.01
Two factors	10.93	25.98
Three factors	11.18	25.92

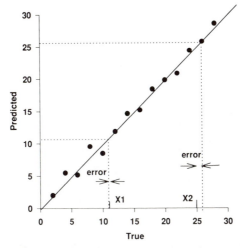

Figure 24 *True vs. predicted tryptophan concentration using a two-factor partial least squares regression model*

As with PCR, a regression model built from two orthogonal new variables serves to provide good predictive ability.

Regression analysis is probably the most popular technique in statistics and data analysis, and commercial software packages will usually provide for multiple linear regression with residuals analysis and variables subset selection. The efficacy of the least squares method is susceptible to outliers, and graphic display of the data is recommended to allow detection of such data. In an attempt to overcome many of the problems associated with ordinary least squares regression, several other calibration and prediction models have been developed and applied. As well as principal components regression and partial least squares regression, *ridge regression* should be noted. Although PCR has been extensively applied in chemometrics it is seldom recommended by statisticians. Ridge regression, on the other hand, is well known and often advocated amongst statisticians but has received little attention in chemometrics. The method artificially reduces the correlation amongst variates by modifying the correlation matrix in a well defined but empirical manner. Details of the method can be found in Afifi and Clark.[6] To date there have been relatively few direct comparisons of the various multivariate regression techniques, although Frank and Friedman[14] and Wold[15] have published a theoretical, statistics based comparison which is recommended to interested readers.

[14] I.E. Frank and J.H. Friedman, *Technometrics*, 1993, **35**, 109.
[15] S. Wold, *Technometrics*, 1993, **35**, 136.

Appendix

A.1

Chemometrics is predominantly concerned with multivariate analysis. With any sample we will make many, sometimes hundreds, of measurements in order to characterize the sample. In optical spectrochemical applications these measures are likely to comprise absorbance, transmission, or reflection metrics made at discrete wavelengths in a spectral region. In order to handle and manipulate such large sets of data, the use of matrix representation is not only inevitable but also desirable.

A *matrix* is a two-way table of values usually arranged so that each row represents a distinct sample or object and each column contains metric values describing the samples. Table 1(a) shows a small data matrix of 10 samples, the percent transmission values of which are recorded at three wavelengths. Table 1(b) is the matrix of correlations between the wavelength measures. This is a *square matrix* (the number of rows is the same as the number of columns) and it is *symmetric* about the *main diagonal*. The matrix in Table 1(c) of the mean transmission values has only one row and is referred to as a *row vector*. This vector can be thought of in geometric terms as representing a point in three-dimensional space defined by the three wavelength axes, as shown in Figure 1.

Matrix operations enable us to manipulate arrays of data as single entities without detailing each operation on each individual value or *element* contained within the matrix. To distinguish a matrix from ordinary single numbers, or *scalars*, the name of the matrix is usually printed in bold face, with capital letters signifying a full matrix and lower-case letters representing vectors or one-dimensional matrices.

Thus if we elect to denote the data matrix from Table 1(a) as A and each row as a vector r and each column as a vector c then,

$$A = \begin{bmatrix} r_1 \\ r_2 \\ \vdots \\ r_m \end{bmatrix} = [c_1 \quad c_2 \quad \ldots \quad c_n] \tag{1}$$

Table 1 *The percent transmission at three wavelengths of 10 solutions, (a), the correlation matrix of transmission values (b), and the mean transmission values as a row vector (c)*

(a)

Sample	λ_1	λ_2	λ_3
	% Transmission		
1	82	58	54
2	76	76	51
3	58	25	87
4	64	54	56
5	25	32	35
6	32	36	54
7	45	54	22
8	56	17	83
9	58	59	62
10	47	65	45

(b)

1.00	0.61	−0.10	0.95	−0.99	−0.74	0.37	−0.03	−0.78	−0.30
0.61	1.00	−0.85	0.33	−0.73	−0.99	0.96	−0.81	−0.97	0.58
−0.10	−0.85	1.00	0.23	0.26	0.74	−0.96	1.00	0.69	−0.92
0.95	0.33	0.23	1.00	−0.88	−0.48	0.06	0.29	−0.54	−0.58
−0.99	−0.73	0.26	−0.88	1.00	0.84	−0.52	0.19	0.87	0.14
−0.74	−0.99	0.74	−0.48	0.84	1.00	−0.90	0.70	1.00	−0.43
0.37	0.96	−0.96	0.06	−0.52	−0.90	1.00	−0.94	−0.87	0.78
−0.03	−0.81	1.00	0.29	0.19	0.70	−0.94	1.00	0.64	−0.95
−0.78	−0.97	0.69	−0.54	0.87	1.00	−0.87	0.64	1.00	−0.36
−0.30	0.58	−0.92	−0.58	0.14	−0.43	0.78	−0.95	−0.36	1.00

(c)

$$r_{\text{means}} = [54.3 \quad 47.6 \quad 45.9]$$

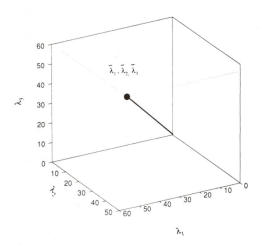

Figure 1 *Vector of means as a point in three-dimensional space*

Table 2 *Some identity matrices*

$$I_2 = \begin{bmatrix} 1 & 0 \\ 0 & 1 \end{bmatrix} \quad I_3 = \begin{bmatrix} 1 & 0 & 0 \\ 0 & 1 & 0 \\ 0 & 0 & 1 \end{bmatrix}$$

where m is the number of rows and n is the number of columns. Each individual element of the matrix is usually written as a_{ij} ($i = 1 \ldots m, j = 1 \ldots n$). If $n = m$ then the matrix is square and if $a_{ij} = a_{ji}$ it is symmetric.

A matrix with all elements equal to zero except those on the main diagonal is called a *diagonal* matrix. An important diagonal matrix commonly encountered in matrix operations is the *unit matrix*, or *identity matrix*, denoted I, in which all the diagonal elements have the value 1, Table 2.

A.2 Simple Matrix Operations

If two matrices, A and B, are said to be equal, then they both must be of the same dimensions, *i.e.* have the same number of rows and columns, and their corresponding elements must be equal. Thus the statement $A = B$ provides a shorthand notation for stating $a_{ij} = b_{ij}$ for all i and all j.

The addition of matrices can only be defined when they are the same size, the result being achieved simply by summing corresponding elements, *i.e.*

$$C = A + B$$

or,

$$c_{ij} = a_{ij} + b_{ij}, \quad \text{for all } i \text{ and } j \tag{2}$$

Subtraction of matrices is defined in a similar way.

When a matrix is rotated such that the columns become rows, and the rows become columns, then the result is the *transpose* of the matrix. This is usually represented as A^{T}. If $B = A^{\mathrm{T}}$ then,

$$b_{ij} = a_{ji}, \quad \text{for all } i \text{ and } j \tag{3}$$

In a similar fashion, the transpose of a row vector is a column vector, and *vice versa*. Note that a symmetric matrix is equal to its transpose.

Matrix operations with scalar quantities is straightforward. To multiply the matrix A by the scalar number k implies multiplying each element of A by k.

$$kA = k.a_{ij}, \quad \text{for all } i \text{ and } j \tag{4}$$

Similarly for division, addition, subtraction, and other operations involving scalar values. The transmission matrix of Table 1(a) can be converted to the

Table 3 *Solution absorbance values at three wavelengths (from Table 1a)*

Sample	λ_1	λ_2 Absorbance	λ_3
1	0.09	0.24	0.27
2	0.12	0.12	0.29
3	0.24	0.60	0.06
4	0.19	0.27	0.25
5	0.60	0.49	0.46
6	0.49	0.44	0.27
7	0.35	0.27	0.66
8	0.25	0.77	0.08
9	0.24	0.23	0.21
10	0.33	0.19	0.35

corresponding matrix of absorbance values by application of the well known Beer's Law relationship,

$$c_{ij} = \log \frac{100}{a_{ij}} \tag{5}$$

with the resultant matrix C given in Table 3.

A.3 Matrix Multiplication

The amino-acids tryptophan and tyrosine exhibit characteristic UV spectra in alkaline solution and each may be determined in the presence of the other by solving a simple pair of simultaneous equations.

$$A_{m,300} = A_{Tr,300} + A_{Ty,300} = \epsilon_{Tr,300} c_{Tr} + \epsilon_{Ty,300} c_{Ty}$$

$$A_{m,200} = A_{Tr,200} + A_{Ty,200} = \epsilon_{Tr,200} c_{Tr} + \epsilon_{Ty,200} c_{Ty} \tag{6}$$

In dilute solution, the total absorbance at 300 nm of the mixture, $A_{m,300}$, is equal to the sum of the absorbance from tryptophan, $A_{Tr,300}$, and tyrosine, $A_{Ty,300}$. These quantities in turn are dependent on the absorption coefficients of the two species, ϵ_{Tr} and ϵ_{Ty}, and their respective concentrations, c_{Tr} and c_{Ty}.

Equation (6) can be expressed in matrix notation as

$$A = \epsilon . C \tag{7}$$

where A is the matrix of absorbance values for the mixtures, ϵ the matrix of absorption coefficients, and C the matrix of concentrations. The right-hand side of Equation (7) involves the multiplication of two matrices, and the equation can be written as

$$\begin{bmatrix} A_{m,300} \\ A_{m,200} \end{bmatrix} = \begin{bmatrix} \epsilon_{\text{Tr},300} & \epsilon_{\text{Ty},300} \\ \epsilon_{\text{Tr},200} & \epsilon_{\text{Ty},200} \end{bmatrix} \begin{bmatrix} c_{\text{Tr}} \\ c_{\text{Ty}} \end{bmatrix} \tag{8}$$

The 2×1 matrix of concentrations, multiplied by the 2×2 matrix of absorption coefficients, results in a 2×1 matrix of mixture absorbance values.

The general rule for matrix multiplication is, if A is a matrix of m rows and n columns and B is of n rows and p columns then the product $A.B$ is a matrix, C, of m rows and p columns:

$$c_{ij} = a_{i1}.b_{1j} + a_{i2}.b_{2j} + \ldots + a_{in}.b_{nj} \tag{9}$$

This product is only defined if B has the same number of rows as A has columns. Although $A.B$ may be defined, $B.A$ may not be defined at all. Even when $A.B$ and $B.A$ are possible, they will in general be different, *i.e.* matrix multiplication is *non-commutative*. If A is a 3×2 matrix and B is a 2×3, then $A.B$ is 3×3 but $B.A$ is 2×2.

The effects of pre-multiplying and post-multiplying by a diagonal matrix are of special interest. Suppose A and W are both $m \times m$ matrices and W is diagonal. Then the product $A.W$ is also a $m \times m$ matrix formed by multiplying each column of A by the corresponding diagonal element of W. $W.A$ is also $m \times m$ but now its rows are multiples of the rows of A. In Table 4(a) the elements of the matrix W are the reciprocals of the maximum absorbances from each of the 10 samples from Table 3. The product $W.A$, shown in Table 4(b), represents the matrix of spectra now normalized such that each has maximum absorbance of unity.

Table 4 *The diagonal matrix of weights for normalizing the absorbance data (a) and the normalized absorbance data matrix (b)*

(a)
$$W = \begin{bmatrix} 0.60 & 0 & 0 \\ 0 & 0.77 & 0 \\ 0 & 0 & 0.66 \end{bmatrix}$$

(b)

Sample	λ_1	λ_2 Absorbance	λ_3
1	0.14	0.31	0.41
2	0.20	0.15	0.44
3	0.39	0.78	0.09
4	0.32	0.35	0.38
5	1.00	0.64	0.69
6	0.82	0.58	0.41
7	0.58	0.35	1.00
8	0.42	1.00	0.12
9	0.39	0.30	0.32
10	0.54	0.24	0.53

A.4 Sums of Squares and Products

The product obtained by pre-multiplying a column vector by its transpose is a single value, the sum of the squares of the elements.

$$x^T.x = \Sigma(x_i)^2 \tag{10}$$

Geometrically, if the elements of x represent the coordinates of a point, then $(x^T.x)$ is the squared distance of the point from the origin, Figure 2.

If y is a second column vector, of the same size as x, then

$$x^T.y = y^T.x = \Sigma(x_i.y_i) \tag{11}$$

and the result represents the sums of the products of the elements of x and y.

$$\frac{x^T.y}{(x^T.x \cdot y^T.y)^{0.5}} = \cos\theta \tag{12}$$

where θ is the angle between the lines connecting the two points defined by each vector and the origin, Figure 3.

If $x^T.y = 0$ then, from Equation (12), the two vectors are at right angles to each other and are said to be *orthogonal*.

Sums of squares and products are basic operations in statistics and chemometrics. For a data matrix represented by X, the matrix of sums of squares and products is simply $X^T X$. This can be extended to produce a weighted sums of squares and products matrix, C:

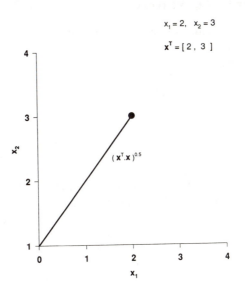

Figure 2 *Sum of squares as a point in space*

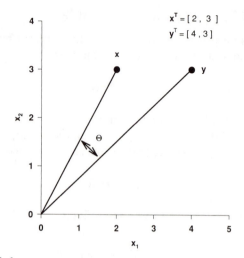

Figure 3   *The angle between two vectors (see text)*

$$C = X^{\mathrm{T}}.W.X \tag{13}$$

where W is a diagonal matrix, the diagonal elements of which are the weights for each sample.

These operations have been employed extensively throughout the text; see, for example, the calculation of covariance and correlation about the mean and the origin developed in Chapter 3.

A.5   Inverse of a Matrix

The division of one scalar value by another can be represented by the product of the first number and the inverse, or reciprocal, of the second. Matrix division is accomplished in a similar fashion, with the inverse of matrix A represented by A^{-1}. Just as the product of a scalar quantity and its inverse is unity, so the product of a square matrix and its inverse is the unit matrix of equivalent size, *i.e.*

$$A.A^{-1} = A^{-1}.A = I \tag{14}$$

The multivariate inverse proves useful in many chemometric algorithms, including the solution of simultaneous equations. In Equation (6) a pair of simultaneous equations were presented in matrix notation, illustrating the multivariate form of Beer's Law. Assuming the mixture absorbances were recorded, and the values for the absorption coefficients obtained from tables or measured from dilute solutions of the pure components, then rearranging Equation (7) leads to

$$\epsilon^{-1}.A = C \tag{15}$$

from which the concentration vector can be calculated.

In general, a square matrix can only be inverted if each of its columns is linearly independent. If this is not the case, and a column is some multiple of another, then the matrix cannot be inverted and it is said to be *ill-conditioned* or *singular*.

The manual calculation of the inverse of a matrix can be illustrated with a 2×2 matrix. For larger matrices the procedure is tedious, the amount of work increasing as the cube of the size of the matrices.

$$A = \begin{bmatrix} a & b \\ c & d \end{bmatrix} \quad \text{and} \quad A^{-1} = \begin{bmatrix} p & q \\ r & s \end{bmatrix} \tag{16}$$

then from Equation (14) we require

$$\begin{bmatrix} p & q \\ r & s \end{bmatrix} \begin{bmatrix} a & b \\ c & d \end{bmatrix} = \begin{bmatrix} 1 & 0 \\ 0 & 1 \end{bmatrix}$$

i.e.

$$\begin{bmatrix} pa + qc & pb + qd \\ ra + sc & rb + sd \end{bmatrix} = \begin{bmatrix} 1 & 0 \\ 0 & 1 \end{bmatrix} \tag{17}$$

Therefore,

$$\begin{aligned} pa + qc &= 1 \\ pb + qd &= 0 \\ ra + sc &= 0 \\ rb + sd &= 1 \end{aligned} \tag{18}$$

Multiplying the first equation by d and the second by c,

$$pad + qcd = d, \quad \text{and} \quad pbc + qdc = 0$$

and subtracting,

$$p(ad - bc) = d$$

or,

$$p = d/(ad - bc) = d/k \tag{19}$$

where $k = (ad - bc)$.

From the second equation we have

$$q = - pb/d = - db/kd = - b/k \tag{20}$$

Similarly from the third and fourth equations,

$$r = -c/k \quad \text{and} \quad s = a/k \tag{21}$$

Thus the inverse matrix is given by

$$A^{-1} = \begin{bmatrix} d/k & -b/k \\ -c/k & a/k \end{bmatrix}, \quad k = ad - bc \tag{22}$$

The quantity k is referred to as the *determinant* of the matrix A, written $|A|$, and for the inverse to exist $|A|$ must not be zero. The matrix $\begin{bmatrix} 1 & 2 \\ 3 & 6 \end{bmatrix}$ has no inverse since $(1 \times 6) - (2 \times 3) = 0$; the columns are linearly dependent and the determinant is zero.

A.6 Simultaneous Equations

We are now in a position where we can solve our two-component, two-wavelength spectrochemical analysis for tryptophan and tyrosine.

A 1 mg l^{-1} solution of tryptophan provides an absorbance of 0.4 at 200 nm and 0.1 at 300 nm, measured in a 1 cm path cell. The corresponding absorbance values, under identical conditions, for tyrosine are 0.1 and 0.3, and for a mixture, 0.63 and 0.57. What is the concentration of tryptophan and tyrosine in the mixture?

From the experimental data,

$$A_m = \begin{bmatrix} 0.63 \\ 0.57 \end{bmatrix}, \quad \epsilon = \begin{bmatrix} 0.4 & 0.1 \\ 0.1 & 0.3 \end{bmatrix} \tag{23}$$

Using Equation (22),

$$|\epsilon| = (0.12 - 0.01) = 0.11 \tag{24}$$

and

$$\epsilon^{-1} = \begin{bmatrix} 0.3/0.11 & -0.1/0.11 \\ -0.1/0.11 & 0.4/0.11 \end{bmatrix} \tag{25}$$

$$\begin{aligned} C_m &= \epsilon^{-1} . A_m \\ &= (1.72 - 0.52), (-0.57 + 2.07) \\ &= (1.2, 1.5) \end{aligned} \tag{26}$$

In the mixture there are 1.2 mg l^{-1} of tryptophan and 1.5 mg l^{-1} of tyrosine.

A.7 Quadratic Forms

To this point our discussions have largely focused on the application of matrices to linear problems associated with simultaneous equations, applications that commonly arise in least-square, multiple regression techniques. One further important function that occurs in multivariate analysis and the analysis of variance is the quadratic form.

The product $x^T.A.x$ is a scalar quantity and is referred to as a quadratic form of x. In statistics and chemometrics A is generally square and symmetric.

If A is a 2×2 matrix,

$$
\begin{aligned}
[x_1 \quad x_2] \begin{bmatrix} a_{11} & a_{12} \\ a_{21} & a_{22} \end{bmatrix} \begin{bmatrix} x_1 \\ x_2 \end{bmatrix} & \\
= (a_{11}x_1 + a_{21}x_2)(a_{12}x_1 + a_{22}x_2) \begin{bmatrix} x_1 \\ x_2 \end{bmatrix} & \\
= a_{11}x_1{}^2 + a_{21}x_1x_2 + x_1x_2a_{12} + a_{22}x_2{}^2 &
\end{aligned}
\tag{27}
$$

and if $a_{21} = a_{12}$ (A is symmetric),

$$
= x_1{}^2 a_{11} + 2x_1 x_2 a_{21} + x_2{}^2 a_{22} \tag{28}
$$

and if $A = I$,

$$
= x_1{}^2 + x_2{}^2 \tag{29}
$$

Thus, the quadratic form generally expands to the quadratic equation describing an ellipse in two dimensions or an ellipsoid, or hyper-ellipsoid, in higher dimensions, as described in Chapter 1.

Subject Index